IX International Conference "Basic Problems of Optics" (BPO 2016)

Journal of Physics: Conference Series
Volume 1062

St. Petersburg, Russia
17 – 21 October 2016

ISBN: 978-1-5108-6829-8
ISSN: 1742-6588

Printed from e-media with permission by:

Curran Associates, Inc.
57 Morehouse Lane
Red Hook, NY 12571

Some format issues inherent in the e-media version may also appear in this print version.

Printed by Curran Associates, Inc. (2018)

For permission requests, please contact the Institute of Physics
at the address below.

Institute of Physics
Dirac House, Temple Back
Bristol BS1 6BE UK

Phone: 44 1 17 929 7481
Fax: 44 1 17 920 0979

techtracking@iop.org

Additional copies of this publication are available from:

Curran Associates, Inc.
57 Morehouse Lane
Red Hook, NY 12571 USA
Phone: 845-758-0400
Fax: 845-758-2633
Email: curran@proceedings.com
Web: www.proceedings.com

Table of contents

Volume 1062

IX International conference "Basic Problems of Optics" BPO'2016

17–21 October 2016, Saint-Petersburg, Russian Federation

Accepted papers received: 6 July 2018
Published online: 23 July 2018

Preface

Papers

New methods for optical data transmission

Applications of Optics in Medicine and Biology

Terahertz Optics and Spectroscopy

Optical Holography

Materials for Optics and Photonics

IOP Conf. Series: Journal of Physics: Conf. Series **1062** (2018) 011001 doi:10.1088/1742-6596/1062/1/011001

The IX International conference "Basic Problems of Optics" BPO'2016

The IX International conference "Basic Problems of Optics" BPO'2016 was held on 17-21 of October, 2016 at ITMO University.

The program of the Conference included series of talks related to different fields of optics, such as quantum optics, optical materials of photonics, optics for biology and medicine, optics of femto- and attosecond pulse. Famous scientists, as well as young specialists, represented their researches in oral and poster presentations.

The topics of the conference were:

- New Methods for Optical Data Transmission
- Application of Optics in Medicine and Biology
- Terahertz Optics and Spectroscopy
- Optical Holography
- Materials for Optics and Photonics

The conference was supported by Optical Society of America (OSA), Society of Photo-optical Instrumentation Engineers (SPIE), Institute of Electrical and Electronic Engineers (IEEE).

IOP Conf. Series: Journal of Physics: Conf. Series **1062** (2018) 011002 doi:10.1088/1742-6596/1062/1/011002

Peer review statement

All papers published in this volume of *Journal of Physics: Conference Series* have been peer reviewed through processes administered by the proceedings Editors. Reviews were conducted by expert referees to the professional and scientific standards expected of a proceedings journal published by IOP Publishing.

IOP Conf. Series: Journal of Physics: Conf. Series **1062** (2018) 012001 doi:10.1088/1742-6596/1062/1/012001

Transceiver for pulsed terahertz spectroscopy based on photoconductive antenna.

S A Korfunenko, Y V Grachev and V G Bespalov

ITMO University, 3b Kadetskaya liniya, St. Petersburg, 199004, Russia

E-mail: skorfunenko@corp.ifmo.ru

Abstract. The paper is concerned with the possibility of creating a pulsed terahertz transceiver. One photoconductive antenna is used for THz radiation generation and detection in this case. The experimental scheme and operation principle of this device are presented.

1. Introduction

Terahertz radiation covers a wide range of electromagnetic waves frequencies: from 0.3 to 10 THz. This radiation has found a wide range of applications, ranging from biomedical to industrial and security systems [1]. Application of terahertz radiation depends on new technical solutions.

Terahertz technology have attracted great interest due to its unique applications in environmental monitoring, space research, chemical identification, material characteristics, the implementations of the inspection, medical imaging, and biological probing [1-3]. At the same time, the practical expediency of many THz systems is still limited by low power, low efficiency and large dimensions of existing THz radiation sources. Photoconductive antennas showed a very promising opportunity among the various methods for THz radiation generation and detection [2-5, 15]. This is mainly due to the existence of compact optical lasers with high efficiency, which allows obtaining a relatively high power terahertz radiation at room temperature using a portable device.

Photoconductive antennas are used widely in terahertz time-domain spectroscopy (TDS) [6-15]. Commonly, THz radiation generation and detection are performed by two individual antennas. Since THz detection is the reverse process of THz generation in photoconductive antenna case, the generator and detector THz radiation can be identical devices operating under different conditions (at constant voltage or under the influence of a THz field).

2. Creation of THz transceiver based on a photoconductive antenna

An apparatus of THz radiation generation and detection by one device is required in modern terahertz time-domain spectroscopy. Such technology will reduce the amount of functional components and reduce economical cost of terahertz spectroscopy. Photoconductive antenna iPCAp-21-05-1000-800-h, produced by Batop optoelectronics used as the generator and the detector of terahertz radiation in created system. Figure 1 show the scheme of the device.

IOP Conf. Series: Journal of Physics: Conf. Series **1062** (2018) 012001 doi:10.1088/1742-6596/1062/1/012001

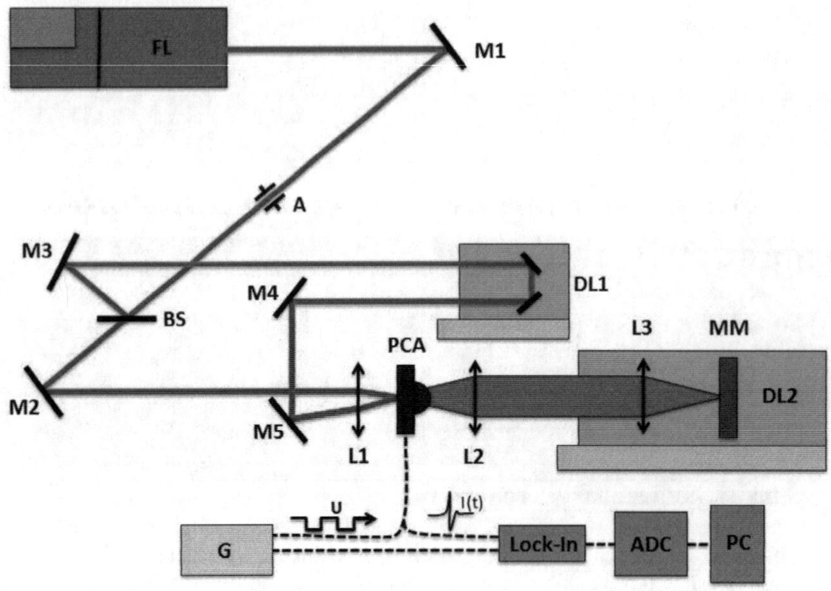

Figure 1. Scheme of THz time-domain transceiver spectrometer.

FL - Ti: sapphire femtosecond laser, A - diaphragm, BS - beam splitter, M1-5 - dielectric mirrors, DL1 - delay line for probe pulse, DL2 - delay line for terahertz pulse, L1 - convex lenses, L2,3 - terahertz lens from TPX, MM - metal mirror, PCA - photoconductive antenna iPCAp-21-05-1000-800-h, G - generator AC and DC voltage, Lock-In - lock-in amplifier LIA-MV-150, ADC - analog-to-digital converter, PC - personal computer PXIe-1062Q

Femtosecond pulses of the pump radiation come to the photoconductive antenna with 64.5 MHz repetition rate which correspond to an oscillation period of 15.5 ns. The pulse energy is about 6.3 nJ. Resistance in the photoconductive antenna is reduced from 31 kΩ to 2.3 kΩ under optical pump. Rectangular gate voltage at a frequency twice lower than femtosecond pulses repetition rate (32.25 MHz) applied to the antenna simultaneously with femtosecond pulses. The bias voltage was varied from 0 to 10 V. The principle of terahertz radiation generation and detection represented graphically at Figure 2.

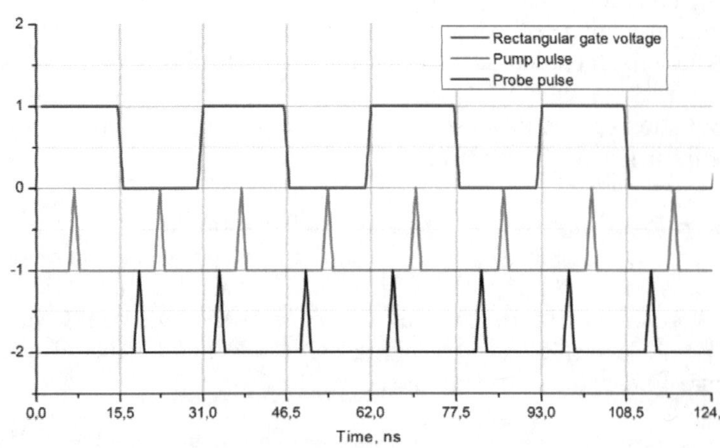

Figure 2. The principle of the generation and detection of terahertz radiation by transceiver.

Rectangular gate AC voltage (blue line), the pump pulses (red line) and offset probe pulses (black line) are shown in Figure 2. Terahertz radiation generation occurred then a voltage is applied to the antenna and femtosecond pump pulses induced charge carriers between photoconductive contact electrodes of antenna PCA. THz pulsed radiation collimated by lens L2 and propagated in free-space to a second lens that focused the terahertz beam on the metallic mirror MM. Terahertz pulsed radiation reflected from the mirror MM and has propagated back to the photoconductive antenna PCA. THz pulse propagation distance was 244 cm, that corresponds to the pulse transit time of about 8 ns. Then terahertz pulse has fell on photoconductor bias voltage hasn't applied to antenna's contact electrodes. Probe beam pulse delayed for the same time as terahertz pulse that allows one to measure photocurrent induced by THz pulse field in the antenna. THz pulse temporary form can be measured by varying the length of the delay line DL1.
Obtained by our terahertz transceiver system THz waveform was presented in Fig. 3.

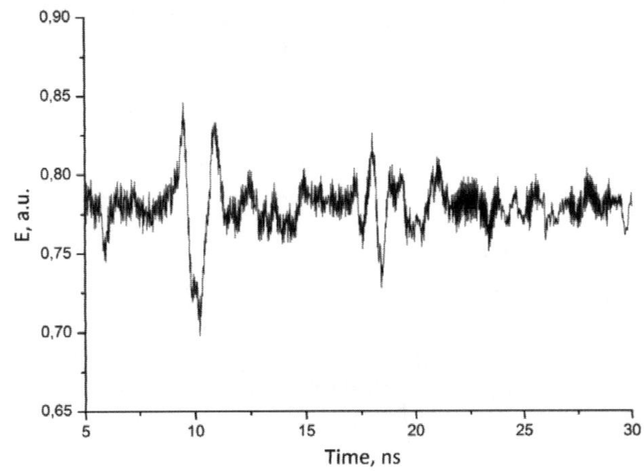

Figure 3. Temporary waveform of THz signal recorded by terahertz transceiver.

Electrical equipment with a time resolution of the order of 1 ns needed to further increase the efficiency of the terahertz transceiver based on photoconductive antenna. According to the results of the research, it is possible to make the conclusion that for the realization THz transceiver it is necessary to synchronize AC voltage repetition rate with the pulse rate of laser system.

3. Conclusion
Ultrashort THz optical pulses are perspective for internal and external integrated circuits (chips) communications for the purpose of time-domain spectroscopy and chemical determination of the composition of complex compounds, for THz radar systems, optical THz imaging purposes nowadays. Device allowing both THz radiation generation and detection significantly reduce the cost, reduce the amount of optical components and simplify construction and adjustment of THz spectroscopy systems.

Acknowledgments
This work was financially supported by Government of Russian Federation, Grant 074-U01.

IOP Conf. Series: Journal of Physics: Conf. Series **1062** (2018) 012001 doi:10.1088/1742-6596/1062/1/012001

References

[1] Zhang X C et al. 2010 *Introduction to THz wave photonics* vol. 29 (New York: Springer)
[2] Tani M et al. 2000 *Electronics letters* ***36*(9)** pp 804–5
[3] Roehle H et al. 2010 *Opt. express* ***18*(3)** pp 2296–2301
[4] Jarrahi M et al. 2008 *Microwave Symposium Digest, 2008 IEEE MTT-S International* pp. 391–94
[5] Grachev Y V et al. 2016 *Infrared, Millimeter, and Terahertz waves (IRMMW-THz) 2016 41st International Conf.* pp. 1–2
[6] Smirnov S V et al. 2016 *Journal of Physics: Conf. Series* **735** 012065 (IOP Publishing)
[7] Semenova V A et al. 2016 *Laser Optics (LO), 2016 International Conf.* pp R7–4
[8] Grachev Y V et al. 2014 *Quantum Electronics,* **44(12)** p 1170
[9] Semenova V A et al. 2016 *Journal of Physics: Conf. Series* vol. 735 p 012063 (IOP Publishing)
[10] Grachev Y V et al. 2014 *Asia Comm. and Photonics Conf.* pp AF4A–5
[11] Balbekin N S et al. 2015 *Journal of Physics: Conference Series* vol 584 p 012010 (IOP Publishing)
[12] Grachev Y V et al. 2014 *Asia Communications and Photonics Conference*
[13] Grachev Y V et al. 2014 *Journal of Optical Technology* **81(8)** pp 468–471
[14] Kulya M S et al. 2017 *Journal of Modern Optics* pp 1–6
[15] Yardimci N T et al. 2015 *IEEE Transactions on Terahertz Science and Technology* **5(2)** pp 223–229

IOP Conf. Series: Journal of Physics: Conf. Series **1062** (2018) 012002 doi:10.1088/1742-6596/1062/1/012002

Modeling of self-consistent modes in optical fibers with V=3.8

A Ibragimov[1], A Rubass[1], S Halilov[1], B Sokolenko[1,2], Ya Akimova[1], M Bretsko[1]

[1]Physical-Technical Institute, V.I. Vernadsky Crimean Federal University. Simferopol, 295007, Crimea
[2]Department of Science and Research, V.I. Vernadsky Crimean Federal University. Simferopol, 295007, Crimea

E-mail: server.khalilov.94@mail.ru

Abstract. The results of modeling the intensity and polarization distribution of both its own and the combination mode of the composition of the isotropic fiber with a gradient a refractive index profile having a waveguide parameter V=3.8.

1. Introduction

The main problem of the current fiber-optic communication is low bandwidth. To increase the number of channels in the fiber, the authors [1-2] used a singular beams. Such beams possess an additional third degree of freedom to increase the number of channels in fiber - orbital angular momentum [3], which does not depend on the polarization state and the wavelength of the radiation beam. To summarize and transmit information along the fiber using singular beam, it is necessary that the waveguide fiber were greater than 2.4. Based on the solutions of the characteristic equation, when a waveguide parameter V=3.8 implemented 12 mod that can carry the singularity [4-8]. Therefore, the aim of this work was to simulate the self-field vector distribution and polarized low-mode optical fiber with V=3.8, and combinations thereof.

2. The construction of intensity and vector distribution of the polarization field of the fiber.

Let us consider the theoretical description of the construction of intensity and polarization patterns for fashion LP_{11} below.

2.1. The theoretical description of the construction of intensity

The field LP_{11} is of mode combination is obtained by adding HE_{21} and TM_{01} or TE_{01} mod [9]. Even the x-polarized input LP_{11} fashion has the following form:

$$E_x = F_1(R)\cos\varphi\cos(\delta\beta_{21}z)\exp(i\tilde{\beta}z) \qquad E_y = iF_1(R)\sin\varphi\sin(\delta\beta_{21}z)\exp(i\tilde{\beta}z) \qquad (1)$$

where $\delta\beta_{21}$ is half-difference propagation constants HE_{21} and TM_{01}, $\tilde{\beta}$-sum propagation constants HE_{21} and TM_{01} mod, to simplify the description of the radial functions used the approximation of a parabolic refractive index profile, then

IOP Conf. Series: Journal of Physics: Conf. Series **1062** (2018) 012002 doi:10.1088/1742-6596/1062/1/012002

$$F_1(R) = R^l \exp(-\frac{VR^2}{2}) \tag{2}$$

where $R = r/\rho$, $\beta_{co} = kn_{co}\left\{1-\frac{4\Delta}{V}(2+l-1)\right\}$, m=1, l=1

The intensity distribution in the case of a vector field is equal to zero Stokes parameters: $I \equiv S_0 = |E_x|^2 + |E_e|^2$. Since the basic object-oriented programming languages there is no complete support for complex numbers, the field is conveniently split into real and imaginary parts. For the field LP_{11} of fashion:

$E_{xr} = F_1(R)\cos\phi\cos(\delta\beta_{21}z)$, $E_{yr} = 0$,

$E_{xi} = 0$, $E_{yi} = F_1(R)\sin\phi\sin(\delta\beta_{21}z)$.

Then the intensity can be represented as:

$$I = S_0 = |E_x|^2 + |E_y|^2 = E_{xr}^2 + E_{xi}^2 + E_{yr}^2 + E_{yr}^2 \tag{3}$$

2.2. The theoretical description of the construction of polarization patterns

(b) According to formula (1), the LP_{11} mode combination at each point of cross-section in the fiber is fully polarized. In the General case of an arbitrary point in the Le has an elliptical polarization. As is known, the polarization is characterized by the following parameters: the intensity I, degree of ellipticity of Q and the tilt angle ψ, the greater semi-axis to the x axis of the laboratory coordinate system. The degree of ellipticity is defined as $Q = \pm b/a$, where the sign "+" is taken if the electric vector rotates clockwise in the plane of observation (light got). These parameters are uniquely determined by the Stokes parameters

$$I = S_0 \tag{4}$$

$$\psi = \frac{1}{2}\arctan(\frac{S_2}{S_1}) \tag{5}$$

$$Q = \tan(\frac{1}{2}\arcsin(\frac{S_3}{S_0})) \tag{6}$$

where:

$$S_0 = |E_x|^2 + |E_y|^2 = E_{xr}^2 + E_{xi}^2 + E_{yr}^2 + E_{yi}^2,$$

$$S_1 = |E_x|^2 - |E_y|^2 = E_{xr}^2 + E_{xi}^2 - E_{yr}^2 + E_{yi}^2,$$

$$S_2 = E_x^* E_y + E_x E_y^* = 2(E_{xr}E_{yr} + E_{xi}E_{yi})$$

$$S_3 = i(E_x^* E_y - E_x E_y^*) = 2(E_{xr}E_{yi} - E_{xi}E_{yr})$$

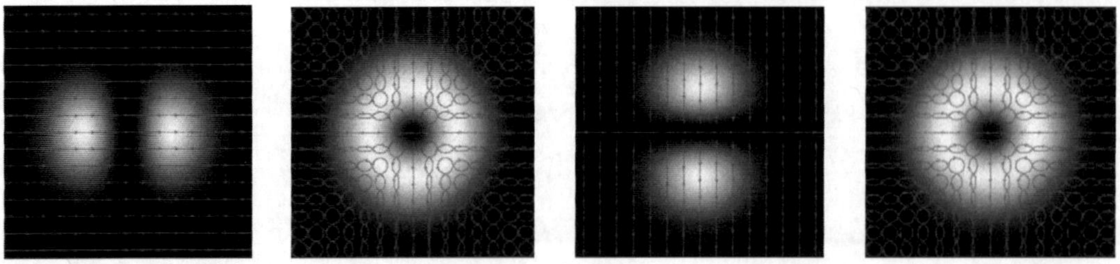

(a) (b) (c) (d)

Figure 1. The distribution of polarization LP_{11} fashion, at λ=650 nm., r=3.5 μm., z=0 cm (a), 45 cm (b), 90 cm (c), 135 cm (d).

IOP Conf. Series: Journal of Physics: Conf. Series **1062** (2018) 012002 doi:10.1088/1742-6596/1062/1/012002

Ellipticity | *The intensity and the distribution of polarization* | *Azimuth*

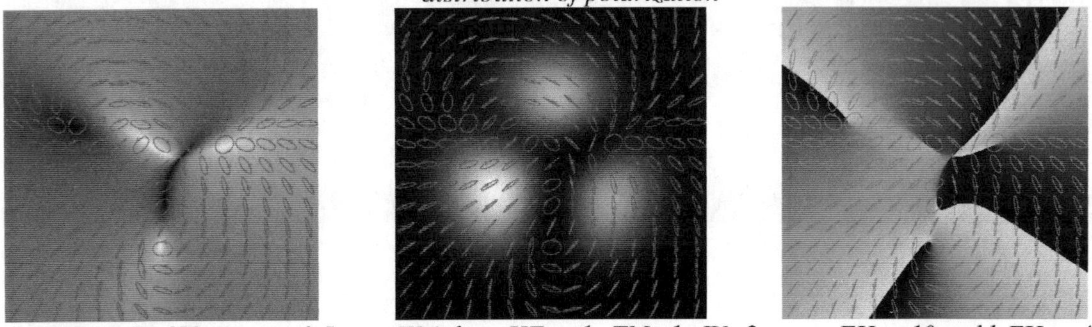

Z=22 cm., λ=650 nm., p=3.5 μm. Weights: HE_{11}=1, TM=1, IV=2, even EH_{11}=10, odd EH_{11}=4, even EH_{21}=3, odd EH_{21}=5

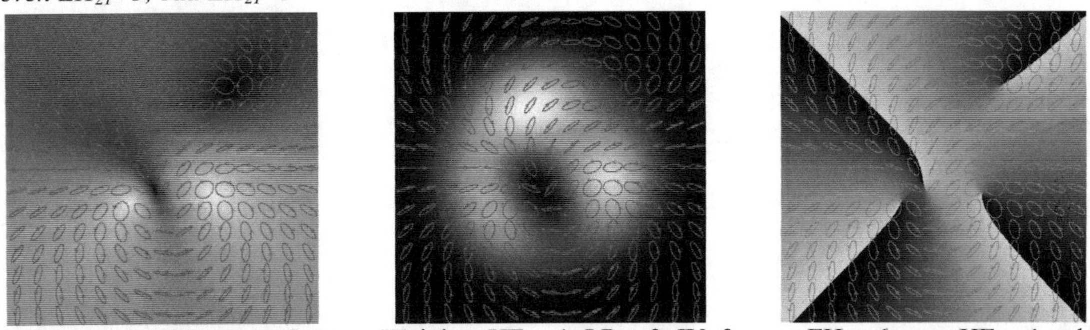

Z=21 cm., λ=650 nm., p=3.5 μm. Weights: HE_{11}=1, LP_{11}=3, IV=2, even EH_{11}=6, even HE_{21}=1, odd HE_{21}=1

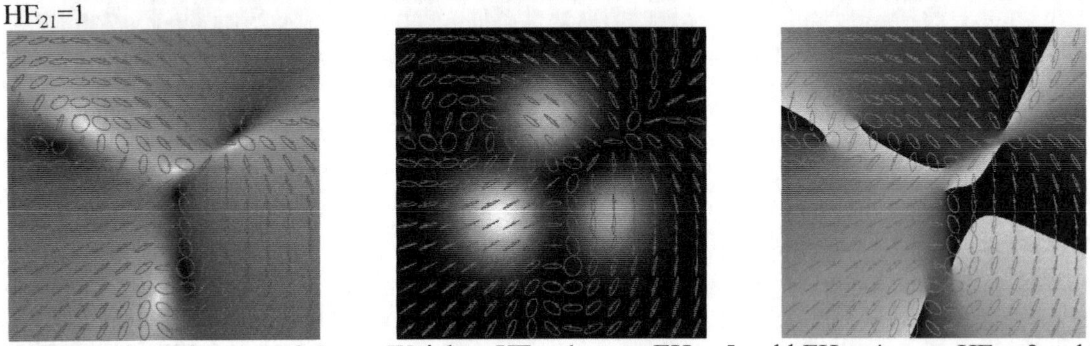

Z=22 cm., λ=650 nm., p=3.5 μm. Weights: HE_{11}=1, even EH_{11}=5, odd EH_{11}=4, even HE_{21}=3, odd HE_{21}=5

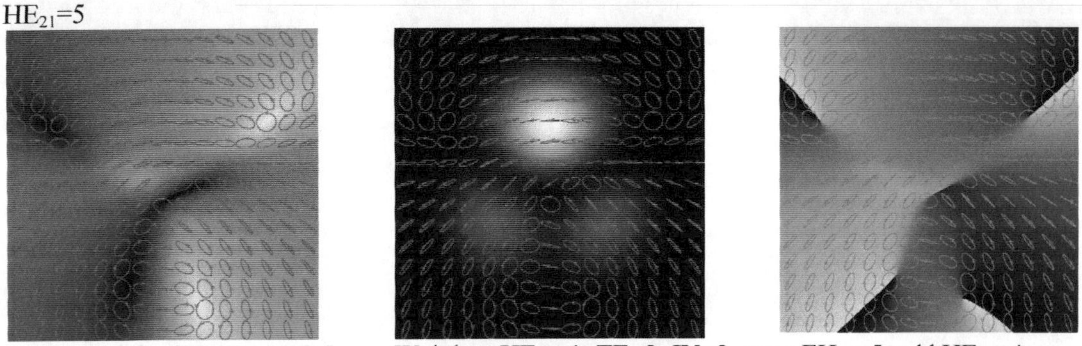

Z=131 cm., λ=650 nm., p=3.5 μm. Weights: HE_{11}=1, TE=3, IV=2, even EH_{11}=5, odd HE_{21}=4.

Figure 2. The superposition of the natural modes of the fiber

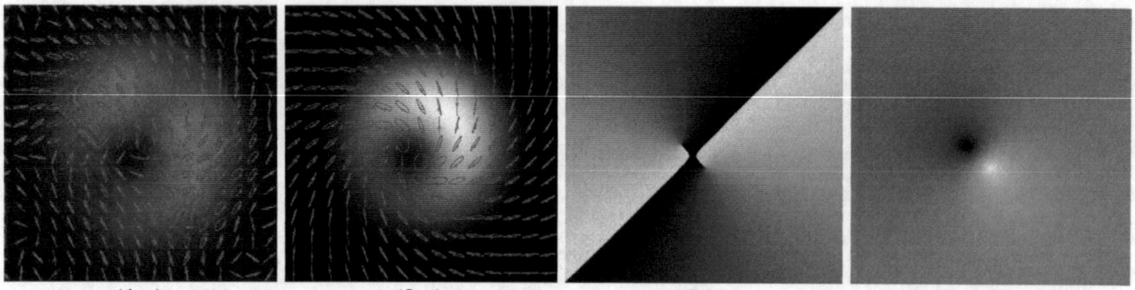

(1.a) (2.a) (2.b) (2.c)

Figure 3. (1) Experiment, (2) modeling: (a) the intensity distribution and polarization, (b) azimuth and (c) ellipticity of the field fiber at Z=20 cm., λ=650 nm., *p*=3.5 μm., with a weight of: *IV=9,* HE_{11}=3

3. Conclusion

In the result, a program was written for simulating basic fashion as well as all possible combinations of the mode composition of propagating in an isotropic fiber with a gradient index profile having a waveguide parameter V=3.8. The program allows you to build a distribution of polarization, the ellipticity and azimuth for the complex modal combinations.

4. Acknowledgement

This work was partially supported by the V.I. Vernadsky Crimean Federal University Development Program for 2015-2024.

References

[1] Bozinovic N, Yue Y, Ren Y, Tur M, Kristensen P, Huang H, Willner A 2013 *Science* **340**

[2] Halilov S, Ilyasova A, Rubass A and Pogrebnaya A 2016 *Journal of Physics: Conference Series* **737**.

[3] Nye J F, Berry M V 1974 *Proc. R. Soc. Lond. A*. **336**. p.165 - 190.

[4] Volyar A V, Zhilatis V Z, Shvedov V G. 1999 *Optics and Spectroscopy* **86** pp. 593-598.

[5] Volyar A V. 2002 *Ukrainian Journal of Physical Optics* **3**

[6] Volyar A V, Fadeeva T A. 1998 *Opt. Spectrosc.* **85** pp. 272-280.

[7] Volyar A V, Fadeeva T A. 1999 *Opt. Spectrosc*. **86**, pp. 242-250.

[8] Volyar A V, Fadeeva T A, Alekseev A N. 2005 *Technical physics letters* **31** pp. 374-377.

[9] Snyder A W, Love J D. Optical 1985 *Waveguide Theory* London: Chapman and Hall – 1985.

IOP Conf. Series: Journal of Physics: Conf. Series **1062** (2018) 012003 doi:10.1088/1742-6596/1062/1/012003

Radiation illusion at the arbitrary location

V Gill, A Vozianova, M Sharaevsky and M Khodzitsky

Dept. of Photonics and Optical Informational Technologies, ITMO University, 49 Kronverkskiy pr., Saint-Petersburg, Russia, 197101

E-mail: viktoria.v.gill@gmail.com

Abstract. The emission illusion device based on metamaterials with radial and angular dependent tensors of permittivity and permeability is analyzed and numerically simulated in this work. The illusion effect was numerically simulated using finite element method for a point source.

1. Introduction

Nowadays, many investigations have been directed to develop variety devices with extraordinary properties such as beam splitters, exotic lenses, cloaks, illusion devices, etc. using transformation optics [1, 2, 3, 4, 5, 6, 7]. The mirage effect creating [8, 9] is of particular interest for communication systems or in order to cheat radar detector.

This work main idea is aimed to virtually modify the spatial location of a emission source. An observer will have the impression that the emitter is located elsewhere instead of its real physical location. The principle of the proposed transformation allows to modify the location of the source as shown in Fig. 1. Indeed, when the emission source with coordinates (ρ_1, φ_1) is placed in the transformed material shell, an observer will see a virtual image of source to appear at another predetermined position (ρ_2, φ_2).

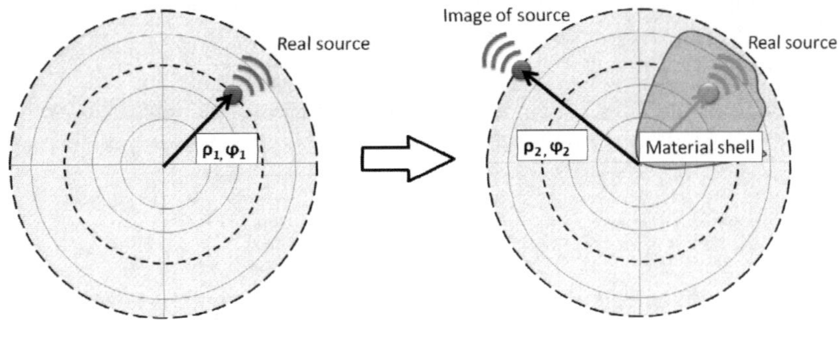

Figure 1. Principle of transformation with virtual delocalization of emission source at the another radial and angular position.

IOP Conf. Series: Journal of Physics: Conf. Series **1062** (2018) 012003 doi:10.1088/1742-6596/1062/1/012003

2. Transformation optics illusion concept and results

Presently the theory to create an illusion has been developed for the cases of preserving the angular component and changing the radial component (Fig. 2) and preserving the radial component and changing the angular component of the radiation source position (Fig. 3).

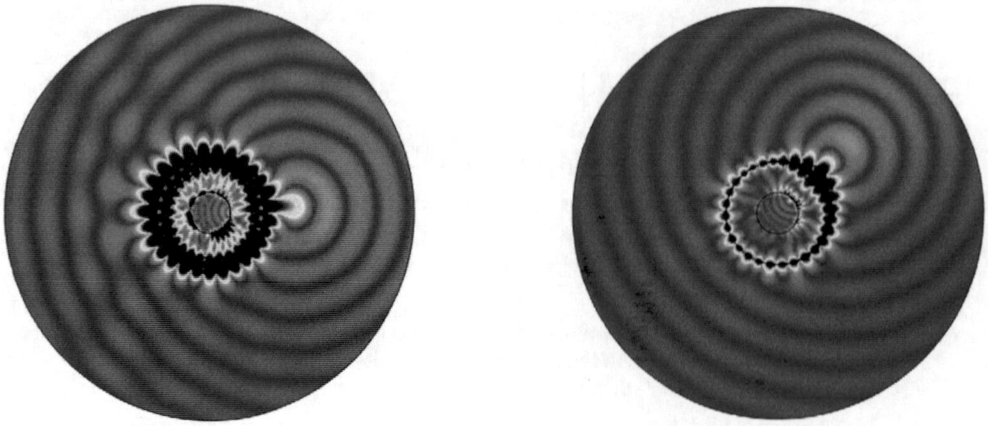

(a) (b)

Figure 2. Simulation case of preserving the angular component and changing (a,b) the radial component of the radiation source position. [10] (5 GHz)

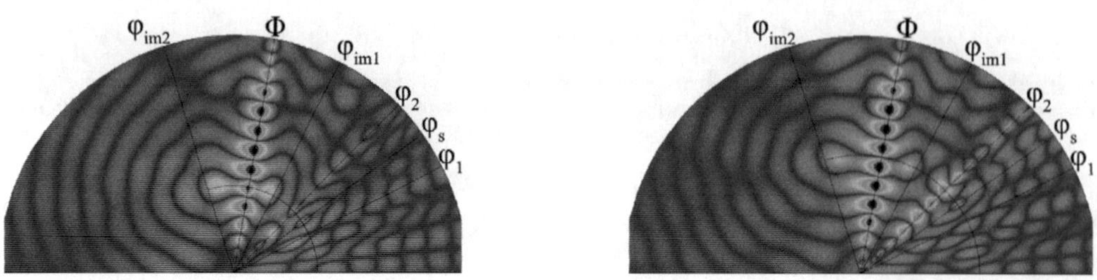

(a) (b)

Figure 3. Simulation case of preserving the radial component and changing (a,b) the angular component of the radiation source position. [11] (5 GHz)

In our work, we create an illusion at an arbitrary point in space with the preset values of the radial and angular component of the position of the source. The transition from the coordinates of the real space to the virtual space is presented below:

$$\rho = \rho(\rho'), \quad \varphi = \varphi(\varphi'), \quad z = z', \tag{1}$$

where $\{\rho, \varphi, z\}$ - coordinates of real space, $\{\rho', \varphi', z'\}$ - coordinates of virtual space.

The dependences of the radial and angular components of the physical space on the corresponding components of the virtual space are presented in equations 2 and 3.

$$\rho = \rho(\rho') = \begin{cases} \dfrac{\rho'}{k}, & 0 < \rho' < R_1 \\ c\rho' + d, & R_1 < \rho' < R_2 \\ \rho', & R_2 < \rho' \end{cases} \tag{2}$$

$$\varphi = \varphi(\varphi') = \begin{cases} \dfrac{\varphi'}{q}, & 0 < \varphi' < \varphi_2 \\ a\varphi' + b, & \varphi_2 < \varphi' < \Phi \\ \varphi', & \Phi < \varphi' \end{cases} \tag{3}$$

where c, d, a, b, k, q - the constants which determine the position of a virtual source.

The permittivity and the permeability tensors are calculated from a well-known expression in terms of Cartesian coordinates as $\varepsilon = \mu = \frac{\mathbf{\Lambda\Lambda^T}}{\mathbf{det(\Lambda)}}\varepsilon'$, here $\varepsilon' = \mathbf{1}$ is the permittivity of homogeneous virtual space and $\Lambda_i^{i'}$ is the transformation matrix from physical polar space $x^i = (\rho, \varphi, z)$ to virtual one $x^{i'} = (\rho', \varphi', z')$.

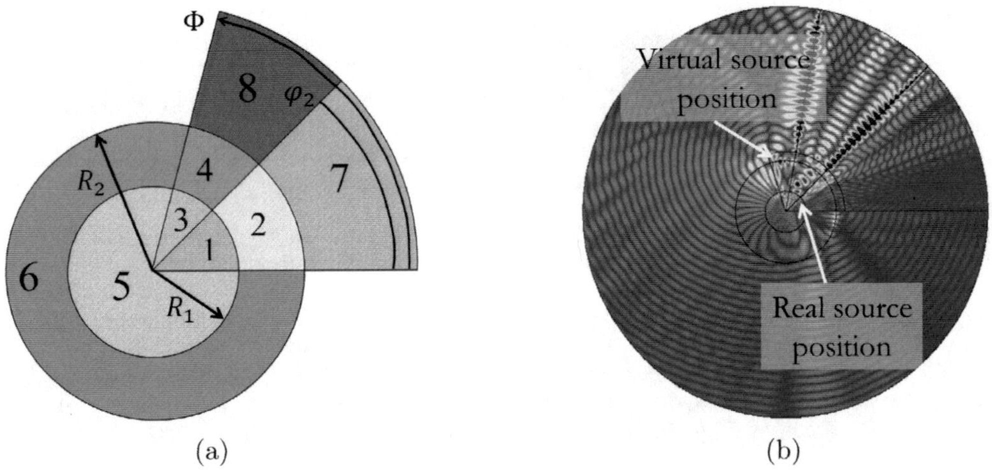

Figure 4. (a) Scheme of designed structure $R_1 = 10\,mm, R_2 = 25\,mm, \varphi_2 = 45°, \Phi = 80°$; (b) Numerical simulation of designed structure at the frequency 40 GHz.

The constitutive parameters of each part of structure of Fig. 4(a) using described above a linear angular transformation is defined as:

$$1.\ \varepsilon = \mu = diag(3, 0.34, 30.72), \tag{4a}$$

$$2.\ \varepsilon = \mu = diag(\frac{3(r-78.6)}{r}, \frac{r}{3(r-0.08)}, \frac{0.65(r-78.6)}{r}), \tag{4b}$$

$$3.\ \varepsilon = \mu = diag(1.6, 0.63, 16.1), \tag{4c}$$

$$4.\ \varepsilon = \mu = diag(\frac{-1.6(r-78.6)}{r}, \frac{-0.63r}{r-78.6}, \frac{-1.6(r-78.6)}{r}), \tag{4d}$$

$$5.\ \varepsilon = \mu = diag(1, 1, 10.24), \tag{4e}$$

$$6.\ \varepsilon = \mu = diag(\frac{r-78.6}{r}, \frac{r}{r-78.6}, \frac{0.28(r-78.6)}{r}), \tag{4f}$$

$$7.\ \varepsilon = \mu = diag(3, 0.33, 3), \tag{4g}$$

$$8.\ \varepsilon = \mu = diag(-1.6, -0.63, -1.6), \tag{4h}$$

where r is measured in $[mm]$.

Using these tensors an optical illusion can be obtained of an arbitrary point in space. Demonstration of the illusion effect is shown in Fig. 4(b) for the case if the source is placed in sector 1.

3. Conclusion

In this work the cylindrical polar radial/angle dependent coordinate transformation has been firstly proposed to create the mirage of a point source radiation at any predetermined position. A linear transformation has been used to obtain constitutive parameters of eight anisotropic materials and to demonstrate an emission illusion. The verification of the analytical principle of such device was performed by numerical simulations using the FEM in the frequency domain. The simulation has demonstrated the lack of influence of the source position on the illusion effect. This idea may be used for dynamical cheating of a radar detection system.

Acknowledgment

This work was supported by Government of Russian Federation (Grant 074-U01).

References

[1] Zentgraf T, Liu Y, Mikkelsen M N, Valentine J, Zhang X *Phys. Rev. Lett.* 2011 **6**(3), 151–155
[2] Kadic M, Guenneau S, Enoch S *Optics Express* 2010 **18**(11), 12027–12032
[3] Gurvitz E A, Sedykh E A, Khodzitskiy M K 2012 *Proceedings of SPIE - The International Society for Optical Engineering*, **8455**, 845532
[4] Gurvitz E A, Vozianova A V, Khodzitsky M K 2014 *Journal of Physics: Conference Series*, **541**(1), 012067
[5] Gurvitz E A, Vozianova A V, Khodzitsky M K 2014 *Proceedings of SPIE - The International Society for Optical Engineering*, **9160**, 91602C
[6] Gill V V, Vozianova A V, Khodzitsky M K 2015 *Journal of Physics: Conference Series*, **643**(1), 012052
[7] Gurvitz E, Andronaki S, Vozianova A, Khodzitsky M *Optics Express* 2015 **23**(20), 25738-25746
[8] Y. Lai, J. Ng, H. Chen, D. Han, J. Xiao, Z.-Q. Zhang, and C. T. Chan, *Phys. Rev. Lett.* , **102**, 253902 (2009).
[9] V.V. Gill, A.V. Vozianova, M.K. Khodzitsky, *Proc. on 10th International Congress on Advanced Electromagnetic Materials in Microwaves and Optics, IET - 2016*, (2016).
[10] J. Yi, P. H. Tichit, S. N. Burokur, A. de Lustracy, *J. Appl. Phys.* , **117**(8), 084903 (2015).
[11] V.V. Gill, A.V. Vozianova, M.K. Khodzitsky, *Book of Abstracts, PIERS2017*, 1121-1122 (2016).

IOP Conf. Series: Journal of Physics: Conf. Series **1062** (2018) 012004 doi:10.1088/1742-6596/1062/1/012004

Optical vortex phase determination for nanoscale imaging

B Sokolenko[1,2*], **N Shostka** [1], **O Karakchieva**[1], **D Poletaev**[2], **V Voytitsky**[3], **S Halilov**[2], **A Prisyazhniuk**[2], **A Ilyasova**[2], **E Kolosenko**[2]

[1]Department of Science and Research, V.I. Vernadsky Crimean Federal University. Simferopol, 295007, Crimea
[2]Institute of Physics and Technology, V.I. Vernadsky Crimean Federal University. Simferopol, 295007, Crimea
[3]Taurida Academy, V.I. Vernadsky Crimean Federal University. Simferopol, 295007, Crimea

E-mail: simplex@crimea.edu

Abstract. In this research we develop a principle of optical vortex phase analysis and its application to surface imaging with high accuracy measurements in nanoscale range. Two-coordinate scanning of the sample allow to retrieve an information about shape and roughness for optically transparent and reflecting surfaces exceeding optical diffraction limit. The interference between singular beam and reference wave, in general, carrying optical vortex with single or doubled topological charge allow to extract the data about phase delay caused by surface features or refraction. This method is also applicable for non-destructive testing of biological structures and live cells in real-time regime. Automatic processing of vortex interferograms allow to achieve a vertical and longitudinal resolution down to 1,75 nm and 7 nm respectively for visible light sources.

1. Introduction

Precise determination of surface roughness and relief is essential task for the manufacturing of optical and mechanical components with high degree of quality and smoothness in engineering, metrology research and materials science applications [1]. For this purposes a great number of instruments and measuring principles were developed last two decades. In general, they are divided into contact and non-contact [2] methods of interaction with the test sample. The first principle uses stylus probe detector and based only on mechanical interaction with an object. This method is inapplicable to the wide range of practical tasks because the scratched surface may be damaged or received parameters are not adequate. Especially it touches such cases as biological systems imaging, polymer structures analysis or metrology of thin film layers.

In contrast, optical methods of profilometry and surface analysis has a lot of advantages, among them: non-contact and non-destructible interaction with test sample [3], relatively high resolution [4], accuracy and flexible control [5-7]. The principle of optical profilometry based on interferometric measurements of optical field, consisting of amplitude and phase information received from the surface being studied [8, 9]. Phase images may be considered as two-dimensional phase distribution or as optical path difference of interfering beams. The relation between geometrical heights $h(x, y)$ of surface profile of optically homogeneous reflecting objects to the phase $\varphi(x, y)$ of reflected wave may be expressed by simple equation [1]:

$$h(\mathrm{x},\mathrm{y}) = \frac{\lambda\varphi(\mathrm{x},\mathrm{y})}{4\pi}, \qquad (1)$$

where λ is the wavelength of interfering light beams.
For imaging of optically inhomogeneous transparent threads, we can assume the dependence of optical path difference is proportional to local projection of refractivity index $n(\mathrm{x},\mathrm{y},\mathrm{z})$ on the beam direction:

$$h(\mathrm{x},\mathrm{y}) = \frac{\lambda\varphi(x,y)}{4\pi}\frac{1}{n(x,y,z)}. \qquad (2)$$

In this technique, the main task is in choice of phase recovering procedure to define phase difference between interfering wave fronts by the values of registered intensity. Temporal modulation, or phase shifting interferometry [10], records sequential interferograms by changing the phase of a reference beam with respect to a sample beam. For this purposes it uses no less than a three interferograms for each point of imaging object [8, 9]. Numerous papers considered methods with using of multi interferograms sequence to extract the phase and surface relief [11]. A profilometer of this type, provides relatively high measurement accuracy of the order of 1 nm, is known, for example, in [12]. The method of multi-step interferograms provides high accuracy of measurements for stationary interference fields, quick data acquisition and large amount of image pixels. The main disadvantage of multi-step methods associated with determination of local phase values in terms of intensity $I(\mathrm{x},\mathrm{y})$, measured during different time intervals. Therefore, fluctuations of light source intensity and frequency as well as vibrations and mechanical instability of interferometric setup impose accuracy limits, especially in dynamic processes and biological applications [13].
The main issues of phase microscopy and imaging are caused by interpretation of collected information from the sample. Phase retrieval algorithm is a practically important, thereby, the modern optical branches, especially singular optics, become an effective instrument in salvation the problem of creation multifunctional microscopes and profilers which use numerical methods in image and data analysis. In numerous manuscripts which describe the methods of surface reconstruction from primary interference distributions, authors used the Fourier [14] and Hilbert transform [13, 15] for phase extraction. Recently, spatial phase-shifting algorithm and derivate method have been developed to enhance the speed of field retrieval without using computationally intensive calculations [16, 17].
Another branch of methods is non-interferometric acquisition can also be used. Non-interferometric analyzers recorded a series of intensity images measured at different axial positions, from which a quantitative phase image is reconstructed with interactive phase retrieval process, which is based on the optical transfer function of intensity equation [18, 19].
Currently the most popular methods include the measurement of light modulation envelope and phase estimation or use its combination. The phase of the electromagnetic wave is determined by using interferometer with considering of refractive index of the material in case of transparent object under investigation. Phase estimation method is the most suitable data processing method in 3D optical microscopy and profilometers with using of both smooth plane waves and beams with phase singularities carrying helical wavefront.
A practical application of singular beams in the vortex scanning optical imaging allows to study, for example, surface geometry and optical density of the sample by analysis of singular phase transformation [20] and depends of the features of incident beam and aperture of optical systems [21]. In manuscripts [22-24] authors demonstrated a new solution for visualization and characterization of nanometer structures called Optical Vortex Scanning Microscope where the sample is scanned by moving vortex. This study demonstrates the response of the optical vortex imbedded in focused Gaussian beam to the shift from the critical plane inside the object arm of interferometer. Scanning procedure enables to cover all area of surface and plot a vortex trajectory, which has a characteristic way of movement and depends on thickness of the probe. Further research of Optical Vortex Scanning Microscope conducted with developing of analytical models and phase retrieval algorithms [25, 26].

In the recent research [27] authors describe both theoretical and experimental results of imaging system using movable optical vortex, where the image of the probe was combined with the structured singular beam. Nevertheless, the phase distribution after the object may be recovered with quite good accuracy, thus using of optical vortices in imaging and microscopy opens new perspectives for development of relief retrieval algorithms and new optical instruments design.
By analyzing of captured interference patterns in form of vortex spiral or retrievable light structures, we can extract an information about phase shift between superposed beams and, as result, about surface roughness state and relief or sample thickness [28] and the task for optimization of such procedures on practice is the purpose of given research.

2. Phase retrieval from interference pattern of coherent waves

Interference of two smooth optical beams, converging at some angle, is often used in modern optical sensors of physical values. Interference pattern of reference and subject wave superposition in Mach-Zehnder or in Michelson interferometer is quite sensitive element and applied to sensors construction with using of fibers, crystals and acousto-optic devices. In certain methods, the physical value is determined by intensity changing within the limited screen area. The disadvantage of this technique is in nonlinear dependence of intensity on the phase difference between beams and allow determining phase difference with low accuracy up to π or using approximately linear area of dependence between intensity and phases in narrow range of phase differences (up to $\pm\pi/2$). It causes on low accuracy of measured value.
Development of a new branch of optics called singular optics [29] allow to create various devices and systems as optical vortex tweezers, beam reducers and sensors, which use singular beams [30-33]. Therefore, it becomes actual to consider the possibility of using singular beams in detectors of physical values, especially in surface roughness measurement and thickness sensing and microscopy [34, 35].
During the interference of singular beam carrying axial optical vortex with the reference Gaussian bean and in case of coaxial spreading with different radii of wavefront curvature, we can observe spiral interference pattern, which rotational angle is determined by phase difference between the beams. This dependence has linear nature, thus it is possible to measure physical value by rotation of interference spiral pattern. At the same time, accuracy of phase determination increases substantially.
In fact, we may use two types of measuring schemes: when singular beam is a probe beam and when vortex imbedded into a reference beam. In both cases, interference patterns are same, but direction of spiral rotation will be inverted.
Therefore, we can conclude that both sensor schemes may be used to determine physical value by the phase interference spiral. The design and alignment of the second sensor scheme is simpler than the first one, but has limitation for our purposes of surface imaging, which will be discussed further.

Let us consider intensity distribution of interference pattern of smooth Gaussian and singular beams. It is defined by the following equation:

$$I = I_1 + I_2 + 2\sqrt{I_1 I_2}\cos(\Phi_1 - \Phi_2), \tag{3}$$

where I_1, I_2 – are the intensities of Gaussian beam and singular beam respectively. Φ_1, Φ_2 – are phases. During co-axial propagation, if beam radii of curvature are much different, an interference pattern is formed on the screen in form of spiral. The position of lightest curve is determined by the value of phase difference:

$$\Phi_1 - \Phi_2 = 2m\pi, \quad (m = 0,1,2K) \tag{4}$$

(for dark curve the phase difference is $\Phi_1 - \Phi_2 = (2m+1)\pi$). The phase of a Gaussian beam is defined as:

$$\Phi_1 = kz + \arctan\left(z/_{z_{0g}} \right) + \frac{kr^2}{R_g(z)} + \varphi_{og}, \tag{5}$$

where $R_g(z) = z\left(1+\frac{z_{0g}^2}{z^2}\right)$ – is the radius of wavefront curvature of Gaussian beam, $z_{0g} = \frac{k\omega_{0g}^2}{2}$ – is the Rayleigh length, ω_{0g} – is the Gaussian beam waist, φ_{og} – denotes initial beam phase. Optical vortex phase is determined by equation:

$$\Phi_2 = kz + (l+1)\arctan\left(z/_{z_{0v}} \right) + \frac{kr^2}{R_v(z)} + \varphi_{0v} - l\phi, \tag{6}$$

where $R_v(z) = z\left(1+\frac{z_{0v}^2}{z^2}\right)$ – is the radius of wavefront curvature of vortex beam, $z_{0v} = \frac{k\omega_v^2}{2}$ – is the Rayleigh length for the singular beam, ω_v, φ_{0v} – are vortex beam waist and initial beam phase respectively, l – is the number of intertwined helices of wavefront or azimuthal index , r – denote polar screen coordinates and z – is the distance to the screen. Phase difference from equations (3) and (4) can be find as follows:

$$\Phi_1 - \Phi_2 = \arctan\left(z/_{z_{0g}} \right) - (l+1)\arctan\left(z/_{z_{0v}} \right) + kr^2\left(\frac{1}{R_g(z)} - \frac{1}{R_v(z)}\right) + \varphi_0 + l\phi. \tag{7}$$

Because on the screen plane $z = const$, phase difference can be expressed at the form:

$$\Delta\Phi = \Phi_1 - \Phi_2 = -ar^2 + b' + l\phi, \tag{8}$$

where $a = k\left(\frac{1}{R_v(z)} - \frac{1}{R_g(z)}\right)$, $b' = \arctan\left(z/_{z_{0g}} \right) - (l+1)\arctan\left(z/_{z_{0v}} \right) + \varphi_0$.

The curve of equal phase on the spiral can be expressed as given in the following equation: $l\phi = ar^2 + b_0$ ($b_0 = \Phi_2 - \Phi_1 + b'$). The simplest case for spiral analysis rises when topological charge of optical vortex $l = 1$. Thus, spiral equation can be expressed as follows:

$$\phi = ar^2 + b_0, \tag{9}$$

Let us consider the influence of geometrical path difference to the phase difference between singular beam and reference arm. Note that vortex phase in probe beam is changing linearly during the variation of path. Total phase shift which is observable due to angular rotation of interference spiral can be calculated from simple equation:

$$\Delta\Phi = \frac{4\pi}{\lambda} n(d_l - d_2) = \frac{4\pi}{\lambda}\mathrm{nh}, \tag{10}$$

where $d_l - d_2 = h$ is a difference between currently observable and neighboring heights of sample surface, which is reflecting the probe beam and n, in general case, is the refractivity index of the media [1, 36]. Therefore, the recorded spiral can be considered as rotational curve (line of constant phase) turned on the angle $\Delta\Phi$ and equation of phase spiral can be expressed as given below:

$$\phi = ar^2 + b_0 + \Delta\Phi, \text{ or } \phi = ar^2 + b. \tag{11}$$

There are several ways to determine the phase difference change of interfering beams [14, 22-25]. We restrict ourselves on finding of intensity minima along spiral phase curve in interference pattern of two waves. In this method, intensity is summed along curvature of equal phase $\phi = const$ (see Eq. 8) as show in Figure 1(c, d). Advantage of this technique is to reduce the noise threshold but still it has strict requirement to accuracy of finding spiral central point. The center of interference spiral according to the equations (6) and (7) coincides with the center of optical vortex accompanied with intensity minima on the beam axis. It can be easily identified by scanning of vortex beam along x and y axis and determination of average intensity minima. As a result, we can find coordinates of spiral center (x_c, y_c) and error of their calculation Δx_c and Δy_c and, finally, roughness state, relief or sample thickness.

3. Singular beams in phase profilometry measurements

In this research we consider analytically an interference of singular beams with the wavelength λ carrying optical vortices of topological charge l. One of the important features of the singular beams is the screw dislocation of wavefront expressed from (8) as beam phase spatial dependence in the form [37]:

$$\Phi(\varphi, z) = kz + l\varphi, \tag{12}$$

where z is the propagation direction, φ is the azimuthal angle at the beam cross section and k denotes wavenumber in a free space. In simplified form, optical vortex with topological charge l may be expressed in terms of Laguerre-Gaussian mode LG_0^l with zero radial index.

Let us consider first the propagation of the paraxial LG_0^l beam along the z-axis. The transverse profile E_x of the beam have a wavenumber $k = nk_o$, where $k_o = \dfrac{2\pi}{\lambda}$ is a wavenumber in a free space and $n = \sqrt{\varepsilon}$ – refractivity index of medium (for convenience it can be assumed $n = 1$). In the paraxial approximation, we can treat the linearly polarised electromagnetic field component as $E = \tilde{E}_x(r,\varphi,z)\exp(-ikz)$ where $\tilde{E}_x(r,\varphi,z)$ is the slowly varying complex amplitude which satisfies the paraxial wave equation [38]:

$$\nabla_\perp^2 E_x(r,\varphi,z) + 2ik\frac{\partial E_x(r,\varphi,z)}{\partial z} = 0. \tag{13}$$

where transverse Laplacian $\nabla_\perp^2 = \dfrac{1}{r}\dfrac{\partial}{\partial r}\left(r\dfrac{\partial}{\partial r}\right) + \dfrac{1}{r^2}\dfrac{\partial^2}{\partial\varphi^2}$, (r,φ) – is the cylindrical coordinates and radius is $r = \sqrt{x^2 + y^2}$. A particular solution to the paraxial wave equation (13) for the Laguerre-Gaussian vortex beam can be written as follows [39, 40]:

$$E_x = \frac{\omega_0}{w}\left(\frac{r}{w}\right)^{|l|} L_m^l\left(m, \frac{r^2}{w^2}\right)\cdot\exp\left(-\frac{r^2}{w^2}\right)\cdot\exp\left(-\frac{ikr^2}{2z(1+z_R^2/z^2)}\right)\times$$
$$\times\exp\left(-il\varphi\right)\cdot\exp\left(i(2m+l-1)\arctan\left(\frac{z}{z_R}\right)\right)\cdot\exp(-ikz), \tag{14}$$

Where ω_0 is the beam waist at the plane in $z = 0$, $z_R = \dfrac{k\omega_0^2}{2}$ denotes Rayleigh length of the beam and current beam radius is expressed as $w = \omega_0\sqrt{1+\left(z/z_R\right)^2}$. Topological charge of vortex beam is

IOP Conf. Series: Journal of Physics: Conf. Series **1062** (2018) 012004 doi:10.1088/1742-6596/1062/1/012004

introduced as $l=\pm1,\pm2,\ldots$ The vortex position, by definition, is described by the equation $\mathrm{Re}\,E_x^{o}(x,y,z)=\mathrm{Im}\,E_x^{o}(x,y,z)=0$. In equation (12) we suppose azimuthal angle is $\varphi=\arg(x+iy)$. Numerical calculation of intensity distribution for Gaussian envelope (corresponding to $l=0$ in equation (14) and intensity of whole vortex beam with $l=1$ and typical axial minima are shown on figure 1 (a, b). The interference and phase patterns in figure 1 (c, d) depicts vortex phase as a helix.

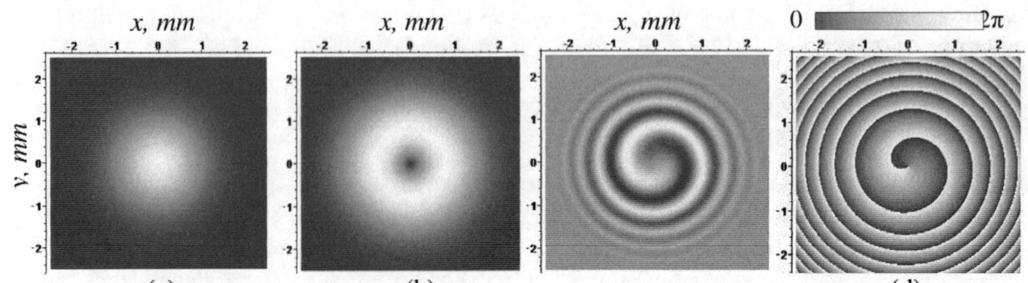

(a) (b) (c) (d)

Figure 1. Theoretically calculated intensity distribution $(\mathrm{Re}\,\mathrm{E}(r,\varphi))^2$ of Gaussian beam (a) when $m=0$ and vortex beam (b) with topological charge $l=+1$ and its interference (c) and phase $\mathrm{Im}\,\mathrm{E}(r,\varphi)$ (d) profile. Other beam parameters are next: $\omega_o=180\,mcm$, $n=1.00$, $z=1m$, $\lambda=632.8\,nm$. Pictures are plotted in Cartesian coordinates.

Singular beam phase is extremely sensitive to changes of optical path between superposed waves. When the optical path of two beams travelling, for example, in arms of interferometer, are equal, it means that waves have the same waist radius and curvature of wavefront. But small variations of optical or geometrical path, which are the same in our case, provokes well-known relative phase shift whether is observable via interference spiral behaviour. Let us consider a case when singular beam propagates through the free space between mirrors of non-equable Mach-Zehnder interferometer and interfere with Gaussian beam. In the object arm we may slightly change the optical path $\Delta=0\div\lambda$, where λ is the wavelength. The resultant phase for different Δ is depicted in figure 2.

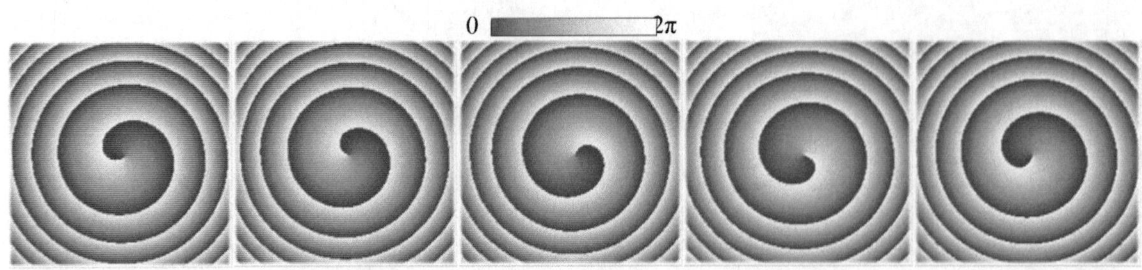

$\Delta=0\,nm$ $\Delta=100\,nm$ $\Delta=250\,nm$ $\Delta=400\,nm$ $\Delta=550\,nm$

Figure 2. Theoretical analysis of singular beam phase $\mathrm{Im}\,\mathrm{E}(r,\varphi,z+\Delta)$ of LG_0^1 beam with topological charge $l=+1$ for different optical path $z+\Delta$. Other beam parameters are next: $\omega_o=180\,mcm$, $n=1.00$, $z=1m$, $\lambda=632.8\,nm$.

Thus, optical path difference manifests itself in phase spiral rotation. The handedness of turn depends of vortex charge sign, in case $l=+1$ it acts on clockwise twist of helical wavefront when Δ increases. Decreasing of path difference on value Δ causes anticlockwise rotation of vortex phase relatively to the reference beam.

IOP Conf. Series: Journal of Physics: Conf. Series **1062** (2018) 012004 doi:10.1088/1742-6596/1062/1/012004

This method [28] makes possible to match phase image or interference pattern with real parameters as local thickness of thin nanometer film or geometrical path difference caused by surface relief. Observable on practice interference fringes or spiral need further image processing [23-25] for phase retrieval, where wide range of algorithms allow to extract φ of the beam with high accuracy [14].

The simplest way for rapid and computationally non intensive analysis is using of direct data from image sensors and cameras with minimal adjustment and transformations. For this purpose we propose to employ superposition of high sensitive probe beam carrying optical vortex with the reference one, but also containing singularity with opposite sign [38]. Phase perturbations caused by geometrical path difference between the beams will noticeably reflect on intensity of resulting image. To understand this, let us consider interference of singular beams described by equation (14) with opposite topological charges $l = \pm 1$ and $l = \pm 2$. The numerical simulation of this process is shown in figure 3.

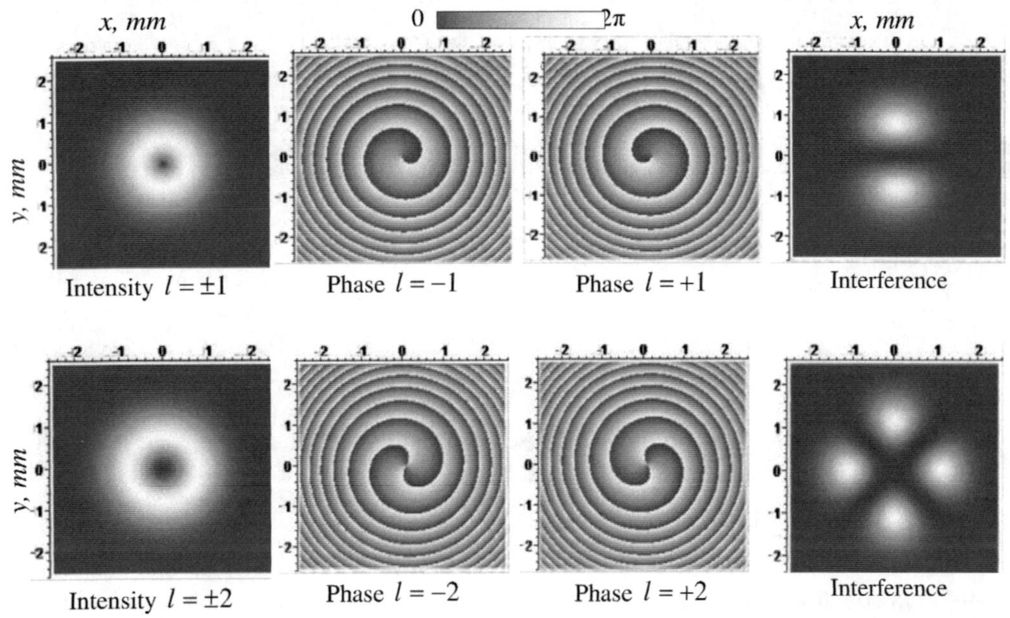

Figure 3. Numerical model of superposition of two optical vortices with opposite signs. Intensity distribution $(\mathrm{Re\,E}(r,\varphi))^2$ of singular beam (first column) when $l = \pm 1$ in first row and $l = \pm 2$ in the second row. Phase portraits $\mathrm{Im\,E}(r,\varphi)$ of vortices with opposite signs are shown in second and third rows respectively. An Interference of both vortex beams is depicted in fourth column for each topological charge. Other beam parameters are the next: $\omega_o = 180\ mcm$, $n = 1.00$, $z = 1m$, $\lambda = 632.8\ nm$. Pictures are plotted in Cartesian coordinates.

Applying the same methodology and reasoning as in case of vortex phase rotation in dependence of different geometrical path difference, we can show dynamics of superposed singular beams with opposite topological charges when one of them experienced a phase shift. This process in sequential form is shown in figure 4.

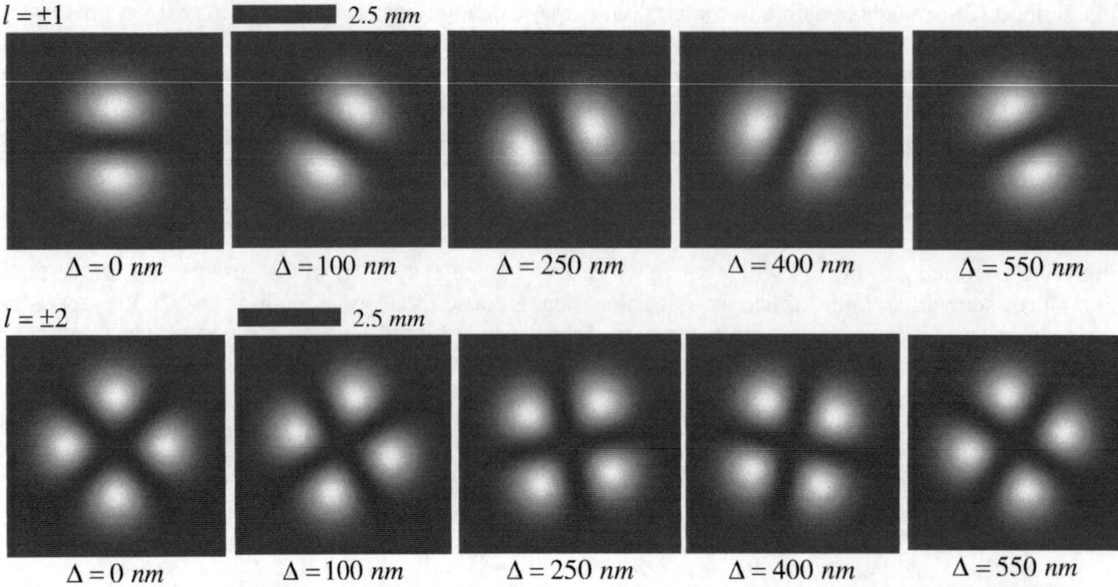

Figure 4. Evolution of superposed vortex beams intensities with topological charge $l = \pm 1$ (first row) and $l = \pm 2$ (first row) for different optical path $z + \Delta$. Other beam parameters are next: $\omega_o = 180\ mcm$, n =1.00, $z = 1m$, $\lambda = 632.8\ nm$.

Making one full turn, intensity pattern coincides with itself at geometrical path difference equal to wavelength λ. Note, that whole picture coincides only ones for topological charges $l = \pm 1$ and twice for $l = \pm 2$. Thus, the symmetry of the pattern caused by topological charges of interfering singular beams. Further increasing of topological charge induces multiplication of high intensity spots around of image centre and are not essential for our application and we restrict ourselves only on superposition of single charged optical vortices for two reasons: at first, interference pattern of single charged vortices has well defined intensity minimum and only one symmetrical axis thus it may be automatically defined with high accuracy. Secondly, the beam spot in this case has much less radius and allow to make beam focusing more easy due to structural stability of optical vortices with $l = \pm 1$. Therefore this method can be used as addition to direct vortex phase analysis as quick draft regime of profile measurement. On estimated evaluation, phase rotation measurement allow to achieve vertical resolution down to $1{,}75\ nm$ for He-Ne laser source with $\lambda = 632.8\ nm$ and less than 1 nm for blue light laser. Rough measurements with resolution $\approx 4nm$ can be performed by processing of interference pattern formed by two coherent optical vortices with opposite signs.

Practical optical vortices appear as rings of light with dark area on the beam axis. A well-known diffraction limit for Gaussian beams was established as relation of wavelength λ to the aperture of optical system by simple equation for spot diameter: $d = 1.22 f \lambda / D$, where f is the focal distance of lens and D is its aperture diameter. To evaluate radius of the beam carrying optical vortex with topological charge l we use the next relation [41]:

$$R_l = a \frac{\lambda f}{\pi D} \left(1 + \frac{l}{l_0} \right), \tag{15}$$

Where $a = 2.585$, $l_o = 9.80$. As result, for practical using we may restrict only central part of the beam carrying single charged vortex which diameter is smaller than diffraction limit. We have measured analytically and experimentally confirmed the relation between beam diameter and the diameter of dark area which is in 2.79 times smaller as shown in figure 5.

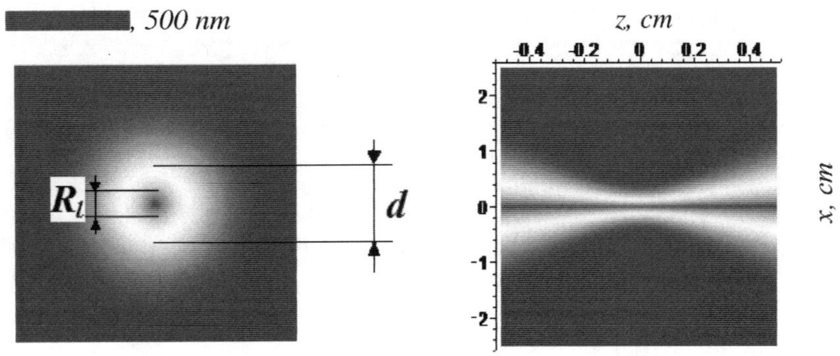

(a) (b)

Figure 5. Schematically depicted tightly focused vortex beam intensity (a) with topological charge $l=+1$ and its transverse spreading in the free space (b). Other beam parameters are next: $\omega_o = 180\, mcm,\ n = 1.00,\ z = 1m,\ \lambda = 632.8\, nm$.

The sharp interference curves with a characteristic "spiral", corresponding to optical vortices can be imaged by the camera and assayed. Total phase shift which is observable due to angular rotation of interference spiral can be calculated from simple equation (10) as a geometrical path difference between observable and neighboring levels of reflecting sample surface. Focused beam spot of 1 mcm on the sample surface perform the longitudinal resolution down to 7 nm.

4. Conclusion

We have analytically considered evaluation of optical phase features and sensitivity to geometrical path changes and have shown that the distinguishable spiral phase rotation occurs at $\lambda/300$, where λ – is a wavelength. Proposed technique may be applied to optically transparent and reflecting surfaces exceed optical diffraction limit. Moreover, this method applicable for non-destructive testing of live cells and biological tissues in real-time regime. Automatic processing of vortex spiral interferograms allow to achieve a vertical resolution down to 1,75 nm. The experimental evaluation of measurement accuracy and finding of optimal phase retrieval procedure is a task of future investigations.

Acknowledgments

This work was partially supported by the V.I. Vernadsky Crimean Federal University Development Program for 2015 – 2024 via grant support for young researches.

5. References

[1] Paul J Caber 1993 *Applied Optics* **32**
[2] Schmit J, Reed J, Novak E, Gimzewski J K 2008 *Journal of Optics A: Pure and Applied Optics* **10**
[3] Zhenrong Z, Jing Z, Peifu G 2010 *Optica Applicata* **40**
[4] Lukin E S, Anufrieva E V, Popova N A, Morozov B A 2009 *Functional Ceramics* **1**
[5] Best G, Amberger R, Baddeley D, Dithmar S, Heintzmann R, Cremer C 2011 *Micron* **42**
[6] Hsieh H, Chen Y, Jian Z, Wu W, Sul D 2009 *J.of Pys. Measurement Science and Technology* **20**
[7] Kitagawa K, Electron J 2012 *Imaging* **21**
[8] Tian C, Liu S 2016 *Optics Express* **24**
[9] Xing S, Guo H 2017 *Applied Optics* **56**
[10] Pham Q D, Hayasaki Y 2015 *Applied Optics* **54**
[11] Tychinsky V P 1989 *Opt. Commun* **74**
[12] Sasaki O, Okazaki H 1986 *Applied Optics* **25**

IOP Conf. Series: Journal of Physics: Conf. Series **1062** (2018) 012004 doi:10.1088/1742-6596/1062/1/012004

[13] Lee K, Kim K, Jung J, Heo J, Cho S, Lee S, Chang G, Jo Y, Park H, Park Y 2013 *Sensors* (Basel) **13**
[14] Takeda T M, Ina H, Kobayashi S 1982 *J. Opt. Soc. Am.* **72**
[15] Ikeda T, Popescu G, Dasari R R, Feld M S 2005 *Opt. Lett.* **30**
[16] Debnath S K, Park Y 2011 *Opt. Lett.* **36**
[17] Bhaduri B, Popescu G 2012 *Opt. Lett.* **37**
[18] Waller L, Kou S S, Sheppard C J, Barbastathis G 2010 *Opt. Express* **18**
[19] Kou S S, Waller L, Barbastathis G, Marquet P, Depeursinge C, Sheppard C J 2011 *Opt. Lett.* **36**
[20] Wang W, Yokozeki T, Ishijima R, Takeda M, Hanson S G, 2006 *Opt. Express* **14**
[21] Popiolek-Masajada A, Sokolenko B, Augustyniak I, Masajada J, Khoroshun A and Bacia M 2014 Optics and Lasers in Engineering **55**
[22] Masajada J, Leniec M, Jankowska E, Thienpont H, Ottevaere H, Gomez V 2008 *Opt. Express* **16**
[23] Masajada J, Leniec M, Augustyniak I 2011 *J. Opt.* **13**
[24] Augustyniak I, Popiołek-Masajada A, Masajada J, Drobczynski S 2012 *Appl. Opt.* **51**
[25] Plociniczak Ł, Popiołek-Masajada A, Szatkowski M, Wojnowski D 2016 *Opt. Las. Technol.* **81**
[26] Plociniczak Ł, Popiołek-Masajada A, Masajada J, Szatkowski M, 2016 *Appl. Opt.* **55**
[27] Popiołek-Masajada A, Masajada J, Kurzynowski P, *Photonics* **4**
[28] Sokolenko B, Poletaev D 2017 Proc. SPIE 10350, Nanoimaging and Nanospectroscopy V
[29] Curtis J E, Grier D G 2003 *Phys. Rev. Lett.* **90**
[30] Curtis J E, Koss B A, Grier D G 2002 *Opt. Comm.* **207**
[31] Friese M E J, Rubinsztein-Dunlop H, Gold J, Hagberg P, Hanstorp D 2001 *Appl. Phys.* Lett. **78**
[32] Paterson L, MacDonald M P, Arlt J, Sibbett W, Bryant P E, Dholakia K 2001 *Science* **292**
[33] Fadeyeva T A, Rubass A F, Sokolenko B V, Volyar A V 2009 *J. Opt. A: Pure Appl. Opt.* **11**
[34] Bowe B W, Toal V 1998 *Optical Engineering* Vol. **37**, Issue 6
[35] Szatkowski M, Popiołek-Masajada A, Masajada J 2014 *Proc. SPIE* **9194**
[36] Wyant J C, Creath K 1992 International Journal of Machine Tools and Manufacture Volume **32**
[37] Bekshaev A, Karamoch A 2009 arXiv *preprint arXiv*:0906.2619
[38] Vickers J, Burch M, Vyas R, Singh S 2008 *J. of the Opt. Soc. of America A*. **25**, Issue 3
[39] Allen L, Beijersbergen M W, Spreeuw R J C, Woerdman J P 1992 *Phys. Rev. A* **45**
[40] Buchhave P, Tidemand-Lichtenberg P 2008 *Opt. Exp.* **16**, Issue 22
[41] Curtis J E, Grier D G 2003 *Optics Letters* **28**, Issue 11

IOP Conf. Series: Journal of Physics: Conf. Series **1062** (2018) 012005 doi:10.1088/1742-6596/1062/1/012005

Diagnostics of biological fluids using dynamic scattering method

T B Kuzmina, N V Andreeva and O V Andreeva

ITMO University, 49 Kronverksky Pr., St. Petersburg, Russia

E-mail: kuzmina.t.94@gmail.com

Abstract. In this paper the research of possibility of using method of dynamic light scattering for early diagnosis of oncological diseases during studying of preparations of biological fluids is presented.

1. Introduction

Currently, because of increasing number of oncological patients also increase interest to the study of oncological diseases, as well as to development of diagnostics and treatment [1]. Timely and proper treatment requires for reducing the percentage of deaths, the effectiveness of which directly depends on the stage at which disease was diagnosed.

Existing methods and programs for diagnostics of oncological diseases based on general blood tests are quite extensive comprehensive program, but mostly they are auxiliary [2]. More and more attention in this area is paid to the development of a broader class of diagnostic methods based on the use of already developed equipment.

The aim of this work is to evaluate the possibility of using the method of dynamic scattering for early detection of cancer during study of biological fluids.

2. Materials and method

The experiment was carried out jointly with the center of heart, blood and endocrinology named after V. A. Almazov. The objects of the research in this work were preparations of whole blood and blood plasma obtained at clinical blood sampling from laboratory mice of two groups: control and experimental (inoculated Erlich carcinoma). Blood for the experiment was taken from laboratory mice of one species, they were kept in the same habitat.

Measurements of distributions of particles by sizes in the preparations of the studied liquids were carried out by analyzer of particles sizes Horiba LB-550.

During this work data on the distribution of particles by sizes in the preparations was obtained. Then the statistical processing of the basic parameters of these distributions was made: there was a comparison of these parameters for control and experimental samples with the aim to identify characteristic changes, by which we can suggest the presence of abnormalities in the sample associated with the emergence of oncological disease. Example of statistical processing is shown in table 1.

Table 1. Example of statistical processing of the parameters of obtained distributions in the preparation of control group.

Sample	№ measurement	1	2	3	4	5	Averaging
control	D_{char}	1.7	2.3	2.6	2.3	2.3	$\langle D \rangle = 2.2 \pm 0.3$
4.1	Q_{max}	28.11	22.22	23.67	24.04	13.79	$Q = 22.37$

3. Results and discussion

During research it was revealed that whole blood has a very large variety of shapes and sizes of the contained components [3], some of which exceed the area of sensitivity of the device.

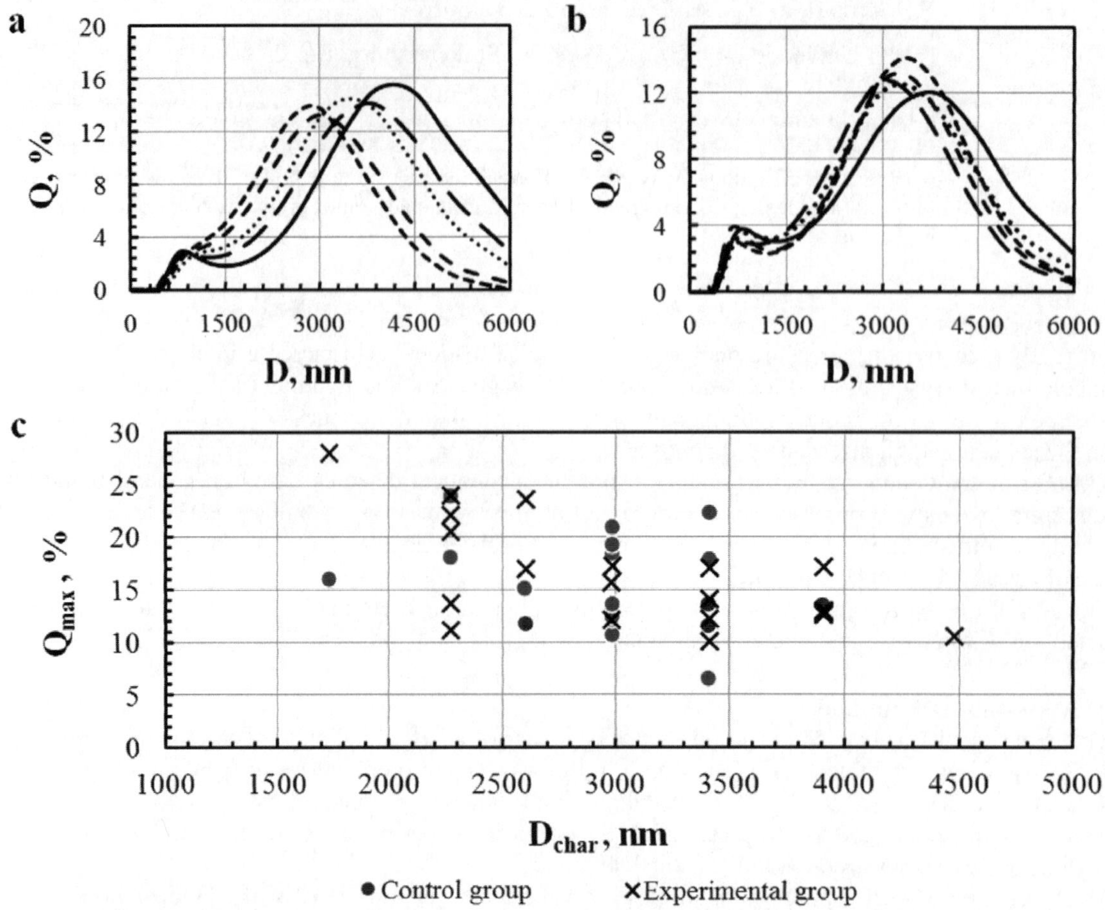

Figure 1. Distribution of particles by sizes of 1% preparations of whole blood: control (a) and experimental (b). The parameters of the distribution function of preparations of whole blood of all samples of two specimen groups (c).

On figure 1 data of distributions of particles by sizes of preparations of whole blood of two specimen groups with concentration of 1% is presented. We can conclude that the character of distributions of particles by sizes doesn't depend on the order of measurements, and accordingly, of the time of beginning of the experiment. In addition, at a constant concentration clearly expressed differences in the resulting distributions couldn't be detected.

Thus, it was decided to pay attention to other components, firstly, the researches with preparations of blood plasma were carried out.
Evaluation of the distribution of particles by sizes in diluted preparations of blood plasma by dynamic light scattering showed the presence of distinct differences of the parameters of the distribution function in preparations of experimental and control specimens.

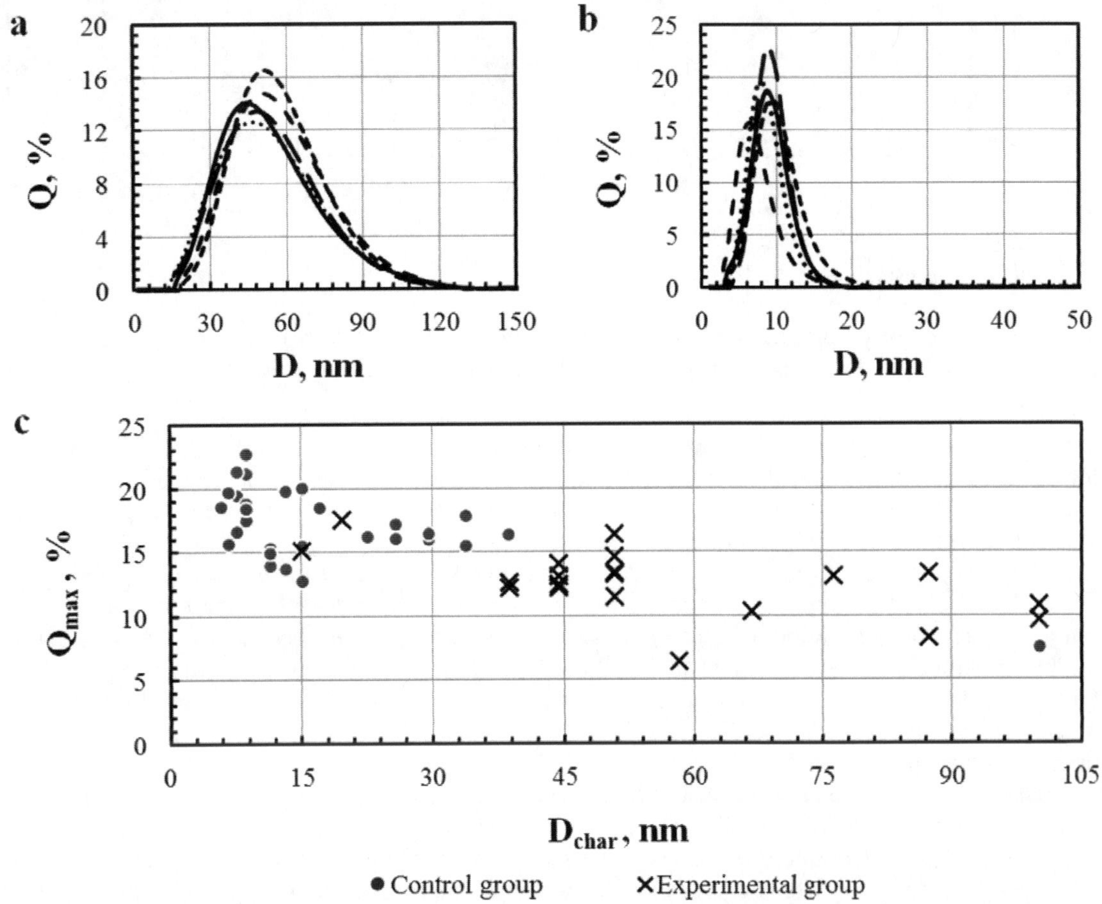

Figure 2. Distribution of particles by sizes of 1% preparations of blood plasma: control (a) and experimental (b). The parameters of the distribution function of preparations of blood plasma of all samples of two specimen groups (c).

In figure 2 data of distributions of particles by sizes of preparations of blood plasma of two specimen groups with a concentration of 1% is presented. According to these graphics, we can notice that the particles in preparations of experimental specimens have larger sizes than the particles of the control group.
The smallest particle size in the researched preparations was 4 nm. In preparations of control group the greatest number of particles have diameter of from 8 to 35 nm. In preparations of experimental group characteristic particle diameter has a value of from 15 to 100 nm, which is much larger than the particles sizes of preparations of control group.

IOP Conf. Series: Journal of Physics: Conf. Series **1062** (2018) 012005 doi:10.1088/1742-6596/1062/1/012005

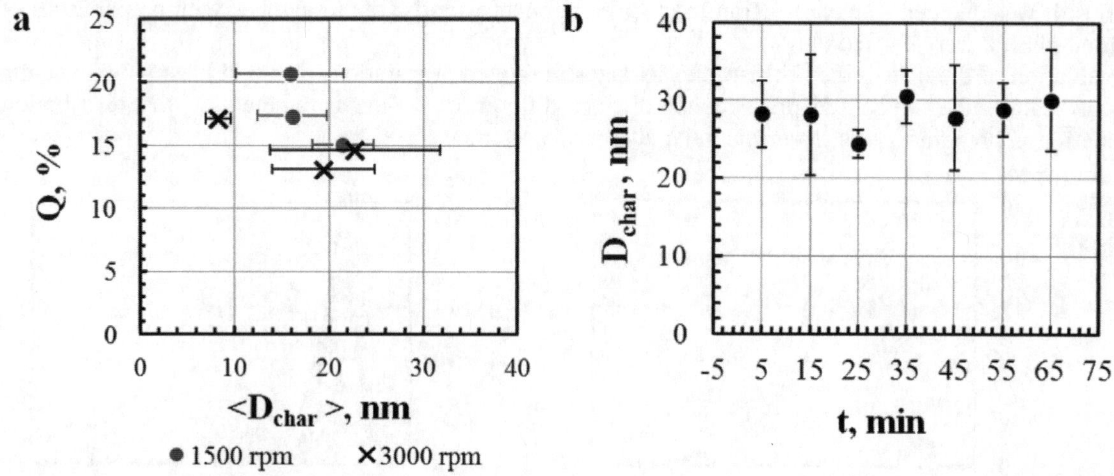

Figure 3. Change of parameters of distribution function of preparations of blood plasma of control group during an hour storage (a). Dependence of the characteristic size of particles in preparation from storage time (b).

In this work the dependence of character of distributions of particles by sizes from the preparation of blood plasma preparations of the control samples was analyzed, it is presented in figure 3(a). For all experiments with blood plasma mode was 1500 rpm. For comparison were used samples of blood plasma obtained at speeds of 3000 rpm.
According to the graphics we can say that at the chosen modes of centrifugation the characteristic size of particles of blood plasma is in the range from 7 to 40 nm, which doesn't contradict the previously obtained values for preparations of blood plasma of control group. Therefore, fixed increase of particle size in preparations of experimental group doesn't stem from the method of sample preparation. However, at higher speeds of centrifugation the statistical spread of obtained values of characteristic size of particles increases. Therefore, mode at 1500 rpm, initially chosen, is more informative for the current experiment.
It was also researched the dependence of characteristic size of particles in the blood plasma of chosen control sample from storage time of preparation, it is shown in figure 3(b). You can see that depending on storage time of preparation, the characteristic size doesn't change within the statistical error. From which it follows that the difference in storage time can't lead to increasing of particles that was fixed in the control group preparations.

4. Conclusion
Obtained at this stage results indicate the possibility of using the method of dynamic light scattering for diagnostics of pathological states of organism associated with the formation of oncological disease. Further development of the method involves the research of other types of carcinomas and tumoral diseases at different stages of their development and at different experimental conditions for achieving the most effective application of the method in the area of early diagnosis of oncological diseases.

References
[1] Torre L A et al. 2015 *CA: a cancer journal for clinicians* **65(2)** pp 87–108
[2] Longo L D 2010 *Harrison's Hematology and Oncology* (McGraw-Hill Medical) p 768
[3] Priezzhev A V, Nikolai N F and Lademann J 2002 *Handbook of optical biomedical diagnostics* ed Tuchin V V (Bellingham: SPIE Press) chapter 11 pp 651–674

IOP Conf. Series: Journal of Physics: Conf. Series **1062** (2018) 012006 doi:10.1088/1742-6596/1062/1/012006

Research of interactions of continuous wave infrared radiation of wavelength 980 nm with mouse tissues

Y D Nazarov[1], Ilatovskaya D.B[1] and M K Khodzitsky[1]

[1] ITMO University, Saint-Petersburg, Russia.

E-mail: yranazarov@yandex.ru

Abstract. In this paper, we used the integrating sphere method to measure the absorption, scattering, attenuation, and scattering anisotropy at 980 nm for the skin, bone and muscle tissue of a mouse.

1. Introduction

In recent decades, the usage of laser radiation sources in medicine has been widely used, both for the diagnosis of diseases, and for the therapy and treatment of a variety of diseases. Examples of such applications are treatments for pigmented lesions, vascular diseases, acne, and some types of skin cancer[1]. In diagnostics, lasers are used for spectral analysis, autofluorescence spectroscopy of some cells, microfluorimetry and many other applications [2]. However, the possibility of applying radiation to biological objects is limited by the depth of penetration into the tissue, which depends on the wavelength [3-4]. Radiation at a wavelength of 980 nm is used in the treatment of certain types of cancer, it is effectively converted by CuS nanoparticles with a diameter of about 130 nm into thermal energy, which is used to treat cancer. [5] This radiation has bactericidal properties, which helps in the treatment of chronic periodontitis [6]. Also, this radiation accelerates wound healing through the excitation of the PDGF and FGF genes, which are responsible for cell renewal and fibroblast growth [7]. Therefore, it becomes necessary to estimate the absorption coefficient of different biological tissues at a given wavelength for the effective treatment of cancer, periodontal disease, wound healing. The aim of the work is to study the interaction of infrared radiation at 980 nm wavelength with biological objects. The problems solved in this paper are development of a technique for measuring optical characteristics (the coefficients of collimated and diffusive transmission, the diffusive reflection coefficient) using an integrating sphere; calculation of optical characteristics: absorption, scattering, attenuation, reduced scattering coefficient and scattering anisotropy factor of mouse tissues (skin, muscle and bone tissue).

2. Experiment description and data processing

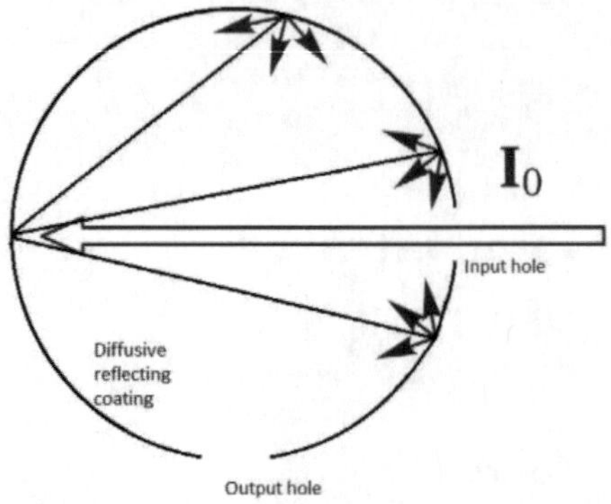

Figure 1. Scheme of the integrating sphere

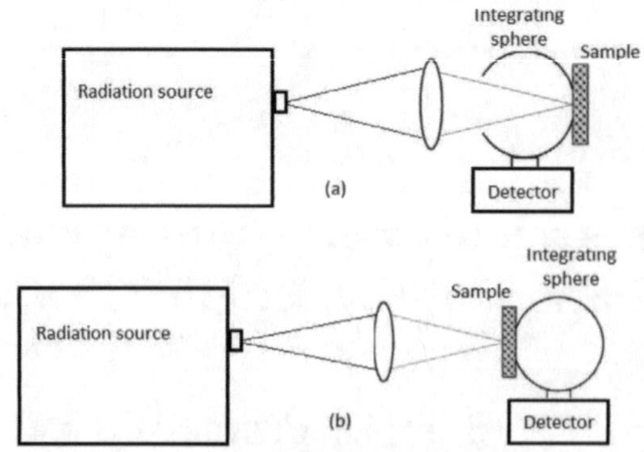

Figure 2. A scheme for measuring the reflection (a) and transmitting (b) of a sample using an integrating sphere.

Since biological tissues and liquids are highly scattering media, the usual methods for measuring the optical characteristics of materials in this case will give distorted results. To measure the parameters of scattering media, the following methods are most often used: laser opto-acoustic method, method of integrating spheres, inverse Monte Carlo method. To carry out this work, a measurement method using an integrating sphere was chosen, because it requires a minimum of additional equipment, is easy to use, but it gives high accuracy. The purpose of the experiment is to obtain a set of optical parameters of the medium at a wavelength of 980 nm. For this experiment, an integrating sphere (figure 1) with an inner diameter of 190 mm, covered from the inside with TiO_2 was created (printed on a 3D printer). The radiation source was a semiconductor laser with a wavelength of 980 nm, power up to 10 W, in a continuous wave mode of generation. To detect radiation, the Ophir Nova II radiation power meter with a 3A-P-SH-V1 sensor was used. During the experiment, the sample was placed in the places indicated in the figure 2, the detector was alternately placed behind the outlet of the radiation source and the integrating sphere to measure the power of the incoming and outgoing radiation. The measurements were repeated 3 to 5 times to reduce the error. The obtained data were processed according to the following formulas:

$$\mathrm{I(Z)} = \mathrm{I_0} \exp(-\mu_t z). \tag{1}$$

The formula (1) used to calculate the attenuation coefficient (μ_t), where Z is the thickness of the sample, I is the radiation intensity after passing the sample, and I_0 is the initial radiation intensity.

$$\mathrm{a} = \frac{1-T_d^2+R_d^2}{2R_d}, \tag{2}$$

$$\mathrm{b} = \sqrt{a^2-1}. \tag{3}$$

The formulas (2) and (3) used to calculate the intermediate variables (a, b), where T_d is the diffusive transmission coefficient, R_d is the diffusive reflection coefficient.

$$S = \frac{1}{bd} \mathrm{Ln}\left[\frac{1-R_d(a-b)}{T_d}\right], \tag{4}$$

$$K = S(a-1). \tag{5}$$

The formulas (4) and (5) used to calculate the Kubelka-Munch absorption coefficient (K) and the Kubelka-Munch scattering coefficient (S), where a and b are the intermediate variables calculated from formulas (2) and (3), d is the thickness of the sample, T_d is the diffusive transmission coefficient, R_d is diffusive reflection coefficient.

$$K = 2\mu_a, \quad (6)$$

$$S = \frac{3}{4}\mu_s(1-g) - \frac{1}{4}\mu_a, \quad (7)$$

$$\mu_t = \mu_a + \mu_s, \quad (8)$$

$$\mu_s' = \mu_s(1-g). \quad (9)$$

Formulas (6), (7), (8) and (9) were used to calculate the absorption coefficient (μ_a), scattering coefficient (μ_s), the reduced scattering coefficient (μ'_s), and the scattering anisotropy factor (g), where K is the Kubelka-Munch absorption coefficient, S is the Kubelka-Munk scattering coefficient, μ_t is the attenuation coefficient. The following formulas were used to calculate the penetration depth (δ), scattering angle (θ), average transport path length of photons (l_δ) and albedo (Λ).

$$\delta = \frac{1}{\mu_a}, \quad (10)$$

$$\theta = \arccos(g), \quad (11)$$

$$l_\delta = (\mu_a + \mu_s')^{-1}, \quad (12)$$

$$\Lambda = \frac{\mu_s}{\mu_t}. \quad (13)$$

3. Results and analysis

Table 1. Obtained optical properties of mouse tissues.

	Skin	Muscle tissue	Bone tissue
μ_t, cm^{-1}	84,9 ± 3,9	28,6 ± 1,4	69,3 ± 3,1
μ_a, cm^{-1}	1,68 ± 0,29	1,44 ± 0,27	2,1 ± 0,37
μ_s, cm^{-1}	83,0 ± 14,6	27,2 ± 5,1	67,2 ± 11,5
μ'_s, cm^{-1}	16,9 ± 3,1	6,67 ± 1,3	16,3 ± 2,9
g	0,80 ± 0,13	0,75 ± 0,12	0,76 ± 0,12

Table 2. Analysis of results.

	Skin	Muscle tissue	Bone tissue
Penetration depth, cm	0,59 ± 0,11	0,69 ± 0,12	0,58 ± 0,10
Scattering angle, degrees	37 ± 7	41 ± 8	40 ± 7
average transport path length of photons, cm	0,05 ± 0,01	0,12 ± 0,03	0,05 ± 0,01
albedo	0,19 ± 0,05	0,23 ± 0,05	0,24 ± 0,05

4. Conclusion

In this paper, we investigated the interaction of infrared radiation at 980 nm with mouse tissues: An integrating sphere was developed and created for measuring the optical characteristics of biological tissues. The coefficients of collimated and diffusive transmission, the coefficient of diffusive reflection of the skin, muscle and bone tissues of the mouse were measured. The absorption, scattering, attenuation coefficients, the reduced scattering coefficient, and the anisotropy factor of mouse tissues were calculated. It was shown that the smallest depth of penetration into bone tissue is 0.58 cm and the largest in muscle tissue is 0.69 cm. It is shown that the smallest scattering angle in the skin is 37 °, the largest in muscle tissue and is 41°. It is shown that the smallest average transport path length of a

IOP Conf. Series: Journal of Physics: Conf. Series **1062** (2018) 012006 doi:10.1088/1742-6596/1062/1/012006

photon in the skin and bone tissue is 0.05 cm, the largest in muscle tissue and is 0.12 cm. It is shown that the smallest albedo in the skin is 0.19, the largest in muscle tissue and is 0.23.

Acknowledgments
This work was supported by Government of Russian Federation (Grant 074-U01)

References

[1] Elma D B *2009 Light-Based Therapies for Skin of Color* (Springler) pp 30–44
[2] Tuchin V V 2010 Lasers and fiber optics in biomedical research (Saratov university) pp 66–91
[3] Duka M V, Dvoretskaya L N, Balbekin N S, Khodzitskii M K, Chivilikhin S A, Smolyanskaya O A 2014 *Quantum Electronics*, ***44***(8), 707-712
[4] Borovkova M, Serebriakova M, Fedorov V, Sedykh E, Vaks V, Lichutin A, Salnikova A, Khodzitsky M 2017 *Biomedical Optics Express 8*(1) 273-280
[5]Dong K, Liu Z, Li Z, Ren J and Qu X 2013 Hydrophobic Anticancer Drug Delivery by a 980 nm Laser-Driven Photothermal Vehicle for Efficient Synergistic Therapy of Cancer Cells In Vivo *Advanced Materials* **32** 4452–4458
[6] Dukić W et al. 2013 Clinical effectiveness of diode laser therapy as an adjunct to non-surgical periodontal treatment: a randomized clinical study *Journal of periodontology* **8** 1111–1117
[7] Usumez A et al. 2014 Effects of laser irradiation at different wavelengths (660, 810, 980, and 1,064 nm) on mucositis in an animal model of wound healing *Lasers in medical science* **6** 1807–1813

Risk zone evaluation for modern technology of varicose veins radiofrequency thermal ablation

D A Rosukhovskii[1,2], E A Iliukhin[3], A A Simonova[1] and M K Khodzitsky[1]

[1] ITMO University, 49 Kronverksky Ave., St.Petersburg, 197101, Russia
[2] Federal State Budgetary Scientific Institution "Institute of Experimental Medicine", 12 Akademika Pavlova Str., St.Petersburg, 197376, Russia
[3] Medalp Vascular Clinic, 54 Leningradskaya Str., St. Petersburg, 197758, Russia

E-mail: rosukhovski@gmail.com

Abstract A widely used radiofrequency ablation technology for varicose veins treatment is accompanied by a high incidence of the adjacent nerves injury in a "risk zone". "Risk zone" term characterizes the region near the vein of the irreversible damage to the nerves and other structures. The mathematical numerical modelling of the propagation of heat emitted by the catheter on the vessel wall was performed using the differential heat equation. It was shown, that at the 20th second of standard radiofrequency ablation cycle the boundary of "risk zone" is located at the distance of 0.38 mm from the vein surface according to the mathematical model of RFA regimen proposed for the first time. The solution injecting under ultrasound surveillance in an amount sufficient to exceed a thickness of 0.38 mm around the vein has been recommended.

1. Introduction

Currently, there is a wide range of thermal ablation technologies for removal of varicose veins (VV). One of the most common of them is a radiofrequency ablation (RFA). Its efficiency for varicose vein removal is up to 97% [1]–[3]. The thermal ablation methods are associated with fewer complications and shorter postoperative period compared to surgery [1], [4]. Nevertheless, an excessive thermal damage which is going beyond VV can injure nearby nerves. Nerve injury is a complication that causes disturbances of sensitivity in up to 30% of cases [5]–[10]. To avoid sensitivity disturbances, VV should be separated from adjacent tissues (including nerves) by a layer of a solution containing anesthetic. The injection is provided by a surgeon with a needle; the solution is pumped along VV under ultrasound surveillance as close to the vein wall as it possible.

The manufacturer of the mostly used equipment for RFA recommends injecting 1 ml of the solution per 1 cm along the vein [3], [11]. However there are no published studies that would establish minimal quantity of this solution or thickness of the solution layer enough for nearby structures protecting during one standard 20-second RFA cycle. The temperature above 60°C is needed for protein coagulation within vein wall, but outside of the vein surface, the same temperature causes nerves injury [12]. "Risk zone" term characterizes the region near the vein of the irreversible nerves and other structures damage. The aim of this study is to calculate the minimum thickness of the solution layer that prevents the nerves injury.

2. Methods

2.1. Preliminary data harvesting

The experimental data of temperature dependences on time on the surface of the catheter placed inside the vein during RFA was taken as initial data in the model. These temperature-time profiles were obtained in a private clinic "Medalp" (St. Petersburg). Video records of 30 RFA procedures from VNUS generator display readings were analysed second by second. The information was received and processed by parametric statistics method. Consequent patients with a diameter of the VV up to 6 mm were included in the research. The mean solution amount that was injected during observed procedures consisted 7,4ml/cm of the vein.

2.2. Geometric model

The VV is considered as a tube with the thickness of 1 mm that tightly covers cylindrical catheter. In our model, a layer of solution evenly surrounds the vein and forms a tube with the infinite thickness around her. The vein and catheter are in isotonic NaCl solution. The geometric model of the studied system is shown in the figure1. All sizes are given in cylindrical coordinates where: $R = 1$ mm – the inner radius of VV/ the outer radius of catheter; $L = 7$ cm – the working length of the catheter; $h = 1$ mm – the wall thickness of the vessel; $B = \infty$ – the thickness of the solution layer.

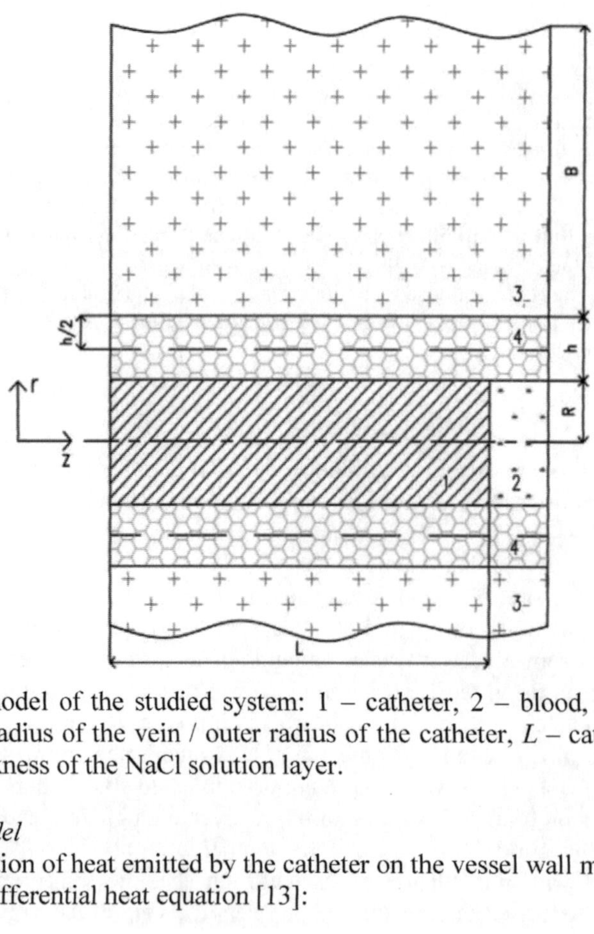

Figure 1. Geometric model of the studied system: 1 – catheter, 2 – blood, 3 – NaCl solution, 4 – vessel wall; *R* – inner radius of the vein / outer radius of the catheter, *L* – catheter length, *h* – vessel wall thickness, *B* – thickness of the NaCl solution layer.

2.3. Mathematical model

The task of the propagation of heat emitted by the catheter on the vessel wall modeling was solved numerically using the differential heat equation [13]:

$$C_p p \frac{\partial T}{\partial t} + \vec{u} \cdot \nabla T - k \cdot \nabla^2 T = Q, \tag{1}$$

here T - temperature; t - time; p, C_p and k - density, specific heat capacity and thermal conductivity, u - velocity of blood in the vessel, Q - energy generated by the catheter during RFA. Assuming that the catheter is fixed, the vein walls are rigid, homogeneous and isotropic, the equation (1) takes the form:

$$C_p p \frac{\partial T}{\partial t} = \frac{\partial}{\partial z}\left(k\frac{\partial T}{\partial z}\right) + \frac{\partial}{\partial r}\left(k\frac{\partial T}{\partial r}\right) + \frac{k}{r}\frac{\partial T}{\partial r} \tag{2}$$

here z and r are longitudinal and radial coordinates of the cylindrical coordinate system. It was assumed, that the values of the parameters used in the model [14], [15] do not have the dependence on the temperature to simplify the modeling. Equations (2) were solved in the 2, 3 and 4 areas (that is, outside the region occupied by the catheter, figure 1) using finite element method [16]. The COMSOL Multiphysics 5.1 software package was used for calculations [17].

2.4. Data interpretation

The so-called "risk zone" term has been introduced in this research to simplify the analysis of RFA procedure safety. So, the "risk zone" defines region near VV at which the temperature reaches more than 60°C during RFA procedure and leads to irreversible damage of nerves and other structures surrounding VV.

The thickness of the solution layer in the case of 1 ml of the solution per 1 cm of the vein introduction, according to recommendations [11] have been calculated. It was assumed that the distribution of the solution around the VV was uniform. For calculations the model with 2 cylinders was used: an internal (venous vessel) cylinder with the radius of 2 mm and external cylinder (volume of the solution) with finite unknown radius x. The difference between the volumes of these cylinders is a recommended solution volume. The height of the cylinders was set equal to 10 mm, that's why, it is possible to operate not with volumes of the cylinders, but squares of cylinders cross-section. That is, the recommended square of cylinders cross-section has been obtained by dividing the recommended solution volume of 1 ml =1000 mm^3 to cylinder height of 10 mm. Thus, the difference between the cross-section of the external and internal cylinders is 100 mm^2 and the radius of external cylinder constitutes 6 mm. Further, the value of the thickness of the solution layer was determined as a difference between the external and internal radiuses of considered system.

3. Results

The recommended thickness of the solution layer has been calculated using the information about the needed solution volume (1 ml per 1 cm of the vein) from the manufacturer guidance. The method of using the difference between the volumes of two cylinders, corresponding to VV and VV with the solution, provides reliable single-valued results due to simplicity of chosen model. Thus the value of solution layer thickness has been constituted 4 mm.

The temperature-time profiles for the surface of the catheter, middle and outer surface of the venous wall were designed for the 20-seconds cycle of the radiofrequency exposure, according to the standard mode of the widely used technology - VeneFit (Figure 2). According to the figure 2 (black line) the vein wall outer surface heating exceeds 60°C at the 9th second of RFA cycle, which can lead to the damage of structures, located in the vicinity of VV and consequently can be cause of the the sensitivity disturbance.

The distance between the vein surface and the isotherm of 60°C near the VV is the "risk zone" depth (Figure 3). At the 20th second of standard RFA cycle the boundary of "risk zone" is bounded above at the distance of 0.38 mm from the vein surface.

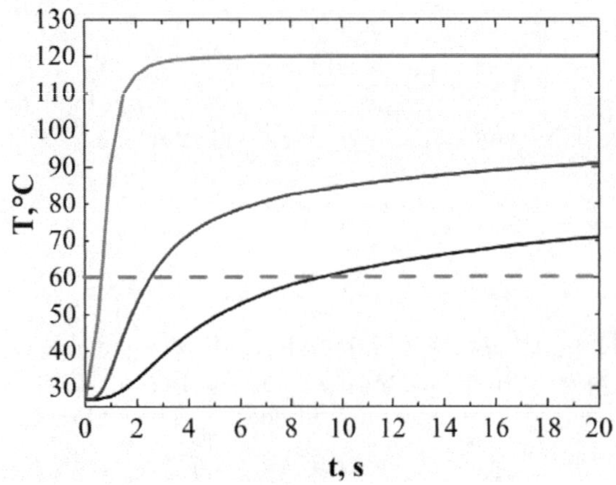

Figure 2. Temperature-time profiles designed as a result of mathematical modeling of standard 20-second cycle of RFA: the catheter surface (red line), the middle (blue curve) and the outer surface (black curve) of the vein wall; the boundary of the effective thermal exposure (T> 60°C) (dotted line).

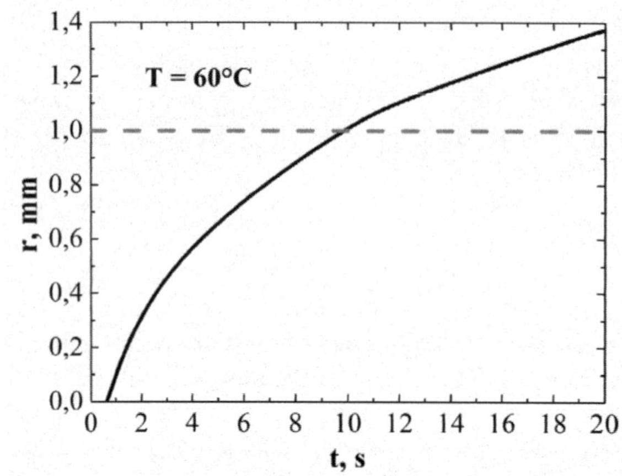

Figure 3. The curves corresponding to the boundary of the "risk zone" for a standard mode (solid black curve) of the catheter with the solution (T = 60°C) and to the surface of venous wall (dotted line).

4. Discussion

We have made the calculations for the thickness of the solution layer, injected around the vein, with an assumption about the uniform distribution of the solution. However, in medical practice, the distribution is uneven. That's why the injection of a larger solution amount is necessary for obtaining a desirable minimum thickness of its layer. Thus, the solution injecting under ultrasound surveillance in an amount sufficient to achieve a thickness of the solution all around the vein not less than 0,38mm is suggested, accordingly to our results.

Further, the thickness of the solution, recommended by the manufacturer, making the simplest mathematical calculation has been defined. This value is 4 mm. Our observation is that surgeons use to inject much more amount of the solution for the lack of the information about the minimal border of so-called “risk zone” for preventing damage to nerves.

The “risk zone” depth obtained in our model is about 10 times less than the solution layer recommended by manufacturer. Our calculations showed that the “risk zone” of nerve injury is relatively thin and can be easily overlapped by the recommended solution quantity. Thus, the nerve injury after the procedure should be explained by quality rather than quantity of the solution injections. It is recommended performing the solution injections between the vein and nerve near the vein wall under the high-resolution ultrasonography.

To avoid the nerve displacement back to the vein due to the leakage of the solution, it is recommended segment-by-segment injecting, starting from the proximal segment immediately before each 7cm of thermal ablation, rather than cover the entire vein in advance as it is accepted to do. In the case, where the distance of 0.38 mm between the vein and nerve cannot be achieved with the help of solution, it is recommended withholding from thermal ablation of this vein segment.

It is possible to eliminate the risk of complications completely by modifying the temperature-time mode of RFA to reduce hyperthermia around the vein to a safe level for nearby tissue. In this case, when the boundary of the "risk zone" aren’t go beyond the vein wall, it is possible to carry out RFA of vein tightly cuddled to the catheter (due to the initially small diameter or spasm, or by increasing the diameter of the working catheter element) entirely without solution under short-term intravenous anaesthesia.

5. Conclusion

It was shown, that at the 20th second of standard radiofrequency ablation cycle the boundary of "risk zone" is located at the distance of 0.38 mm from the vein surface according to the mathematical model of RFA regimen proposed for the first time. The solution injecting under ultrasound surveillance in an amount sufficient to exceed a thickness of 0.38mm around the vein has been recommended.

Acknowledgments

The Government of Russian Federation, Grant 074-U01, supported this work.

References

[1] Wittens C et al. 2015 Editor's Choice - Management of Chronic Venous Disease: Clinical Practice Guidelines of the European Society for Vascular Surgery (ESVS) *Eur. J Vasc. Endovasc. Surg.* **49**(6) 678–737
[2] Treatment of Superficial Venous Disease of the Lower Leg Guidelines 2015 (American College of Phlebology)
[3] Goodyear S J and Nyamekye I K 2015 *Phlebology* **30**(2) 9–17
[4] Varicose veins: diagnosis and management (NICE Clinical guideline) 2016
[5] Choi J H, Park H–C and Joh J H 2013 *J Korean Surg Soc* **84**(2) 107–13
[6] Rasmussen L H, Lawaetz M, Bjoern L, Vennits B, Blemings A and Eklof B 2011 *Br. J. Surg.* **98** 1079–1087
[7] Proebstle T M et al 2011 *J. Vasc. Surg.* **54**(1) 146–152
[8] Tolva V S, Cireni L V, Bianchi P G and. Casana R M 2013 *Surg Today* **43** 741–744
[9] Bisang U, Meier T O, Enzler M, Thalhammer C and Husmann M 2012 *Phlebology*
[10] Park H S, Kwon Y, Eom B W and Lee T 2013 *J Korean Surg Soc* **84** 48–56
[11] Reference Guide For The Venefit Procedure
[12] Sengupta R P and Stunden R J 1977 *Br Med J* **6054** 142–3
[13] Carslaw H S and Jaeger J C 1959 *Conduction of heat in solids* 2nd ed. (London: Oxford University Press)
[14] Choi S Y, Kwak B K and Seo T 2014 *Comput Math Methods Med* 485353
[15] Shadrina E M and Volkova G V 2009 *Determination of thermal properties of gases, liquids and water solutions of substances* (Ivanovo: Ivanovo State University of Chemistry and Technology Publ.)
[16] Jain M K and Wolf P D 2000 *Ann Biomed Eng* **28**(9) 1075–84
[17] COMSOL Multiphysics 2016 (COMSOL Inc)

Monitoring of sedimentation of aqueous suspensions of fine quartz by method of digital holographic interferometry

R N Chigrin, T B Kuzmina, N V Andreeva, O V Andreeva

ITMO University, 49 Kronverksky Pr., Saint-Petersburg, 197101, Russia

E-mail: rchigreen11@gmail.com

Abstract. The usage of the method of digital holographic interferometry (DHI) for the study of fine dispersed suspensions of inorganic substances in the process of their sedimentation was proposed. It is shown that during research of nanosuspensions with particle sizes less than 400 microns method of DHI allows to reveal changes of parameters with greater precision than method of dynamic light scattering (DLS).

1. Introduction

At present time the usage of substances in dispersed state is widely used in biomedical technologies, as well as during creation of new composite materials, to improve the physico-mechanical properties of industrial structural materials [1,2]. The development of nanotechnologies requires the development of methods of diagnostics of nanoobjects parameters. Development of new methodology is an important task as it allows to increase the reliability of the results obtained by other methods [3].

In this work the possibility of characterizing parameters of the nanodispersed inorganic suspensions by monitoring of the process of sedimentation by method of digital holographic interferometry is considered and comparison with the currently used method of dynamic light scattering is carried out [4].

2. Objects and methods

Objects of research in this work are aqueous suspensions of fine quartz SiO_2 which particles were obtained by different methods and were used for different purposes. The particles of suspension 1 are obtained by condensation of silicon dioxide from the gas phase, such particles are used in biomedical technologies and during creation of polymer composite materials for various purposes with improved physico-mechanical properties. The particles of suspension 2 are obtained by grinding of quartz sand in planetary-centrifugal mills and they are used as industrial additives which improve the operational properties of building materials.

Initial state of SiO_2 for production of suspension 1 – dry powder; for production of suspension 2 – pasty mixture with water in a weight ratio of 1.3:1.0. Work preparations were 0.1%-0.5% aqueous suspensions of the researched substances.

For the control of changes of parameters of the researched suspensions in the process of their sedimentation during long time storage two methods were used.

The method of dynamic light scattering was realized on device Horiba LB-550. The research of preparations was carried out in the same 10x10x40 mm cuvettes with a volume of 4 ml. Measurements of distribution of suspensions particles by sizes were carried out at a distance of about 20 mm from the bottom of the cuvette at given time intervals (figure 1c).

IOP Conf. Series: Journal of Physics: Conf. Series **1062** (2018) 012008 doi:10.1088/1742-6596/1062/1/012008

Metod of DHI was realized on experimental stand which was developed for studying phase transformations of transparent objects [5]. The research of preparation of suspension 1 was carried out in 20x50x30 mm cuvette and research of preparation of suspension 2 – in 5x20x30 mm cuvette. The cuvette is rigidly mounted in a special cassette and installed in the object beam in accordance with the optical scheme of interferometric experiment. The object beam of interferometer (532 nm) tests the working area of the object with the size at least 25x25 mm and passes through the sample. The stand has a computer terminal to control a registration of sequence of digital holograms and their further processing.

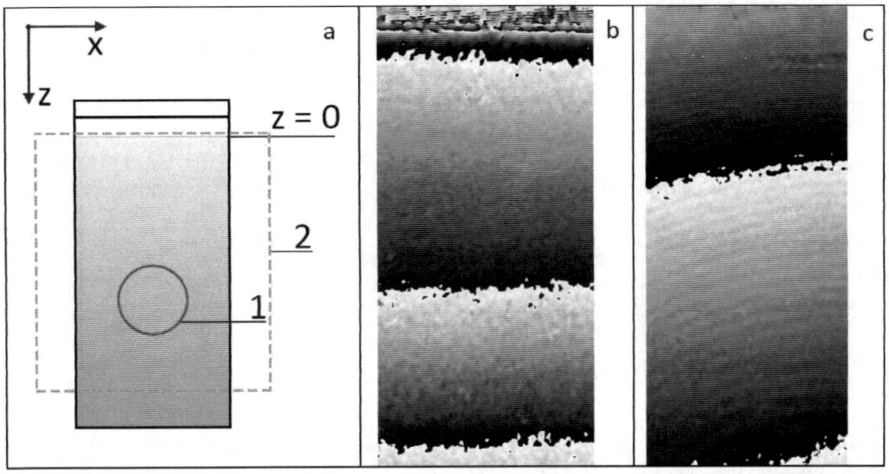

Figure 1. The cuvette with the researched suspension (a): 1 - area of monitoring using method of DLS, 2 – area of monitoring using method of DHI; the form of interferogram (b,c): suspension 1 after 72 hours of sedimentation (b), suspension 2 after 20 hours of sedimentation (c).

The digital interferogram of the current state of the object represents the spatial distribution of the phase difference of the testing radiation in the image plane of the object corresponding to the change of the sample compared to the initial one. The interference pattern has the form of horizontal interference fringes characterize the change of the optical thickness of the sample by height of the cuvette, due to changes of concentration and sizes of particles during their sedimentation (figure 1b,c).

3. Experimental results

Experimental results are shown on figure 2 and in the table.

The distribution of particles by sizes obtained by method of DLS (figure 2a,b) is characterized by the values D_x, where D_x is the typical particle diameter, defined at Q_{max}, and ΔD - half width of distribution of particles by sizes. On figure 2(a,b) the values D_x and ΔD for the studied suspensions are shown, in the initial state (curves 1 - immediately after producing preparation before the process of sedimentation of suspension, $t_s = 0$) and in the process of sedimentation: for suspensions 1 after 20 hours (curve 3) and after 20 days (curve 4); for suspention 2 after 2 hours (curve 2) and 20 hours (curve 3).

On figure 2(c) and 2(d) the results of monitoring of the process of sedimentation of suspensions 1 and 2 by method of DHI are shown. The dependence of the change of $\Delta n(z)$ is given for suspensions in the initial state (curves 1) and in the process of sedimentation (curves 3,4).

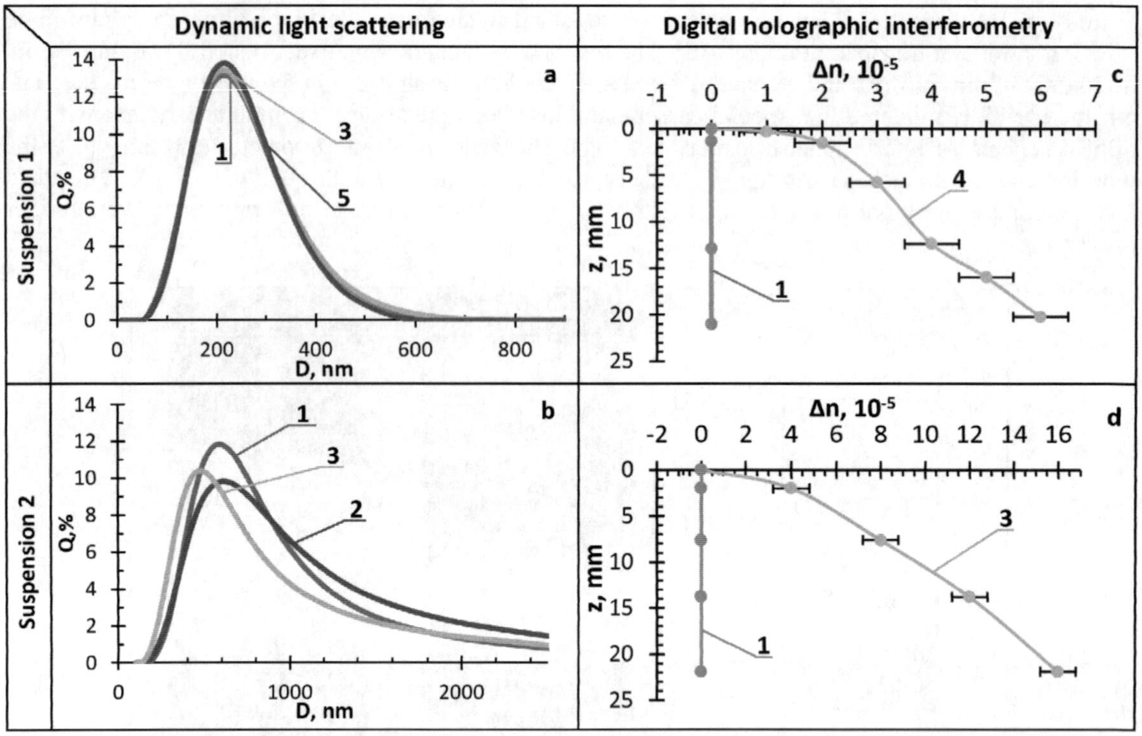

Figure 2. The distribution of particles by sizes (a,b) and the change of the refractive index of suspensions 1 and 2 (c,d) in the initial state (curves 1) and in the process of sedimentation: 2 hours (curve 2), 20 hours (curves 3), 72 hours (curve 4), 20 days (curve 5).

As can be seen, for suspension 1 during sedimentation for 20 days the changes by method of DLS can't be diagnosed, while by method of DHI the changes steadily diagnosed during sedimentation for 72 hours (table).

Table. Estimation of parameters of suspention 1 in the process of sedimentation by methods of DLS and DHI.

Method of DLS				Method of DHI
$t_s = 0$		$t_s = 480$ hours		$t_s = 72$ hours
D_x, nm	ΔD, nm	D_x, nm	ΔD, nm	Δn ($z = 20$ mm)
220	210	220	210	$5.8 \cdot 10^{-5}$

4. Conclusion

The experiments showed that the parameters of suspension 1 with SiO_2 particles (typical particle diameter 220 nm) determined by method of dynamic light scattering practically don't change during 20 days of storage. At the same time method of digital holographic interferometry allows to detect the change of refractive index after 72 hours of suspension storage. The parameters of suspension 2 (typical particle diameter 670 nm) show well-diagnosed changes of parameters by both methods in the first day of storage. Thus, during estimation of parameters of nanodispersed suspensions method of digital holographic interferometry allows to carry out diagnostics with higher accuracy than the method of dynamic light scattering.

IOP Conf. Series: Journal of Physics: Conf. Series **1062** (2018) 012008 doi:10.1088/1742-6596/1062/1/012008

References

[1] Guglielmi M, Kickelbick G and Martucci A 2014 *Advances in Sol–Gel Derived Materials and Technologies* **IX** p 227
[2] Rosenholm J M, Zhang J, Lindén M and Sahlgren C 2016 *Nanomedicine* **11(4)** pp 391–402
[3] Rahimi-Nasarabadi M 2014 *Journal of Nanostructures* **4(2)** pp 211–216
[4] Zheng T, Bott S, Huo Q 2016 *ACS Appl. Mater. Interfaces* **8(33)** pp 21585–94
[5] Manukhin B G, Gusev M E, Kucher D A, Chivilikhin S A and Andreeva O V 2015 *Opt. Spectros.* **119(3)** p 392

Two-photon absorption in THz electro-optical sampling crystals

M O Zhukova[1], E A Makarov[1], S E Putilin[1], A N Tsypkin[1], V P Chegnov[2], O I Chegnova[2] and V G Bespalov[1]

[1] ITMO University, 49 Kronverksky Pr., St. Petersburg, 197101, Russia
[2] Research Institute of Materials Science and Technology, proezd 4806, build. 4/2, Moscow, 124460, Russia

E-mail: zhukovamo@ya.ru

Abstract. In this paper we present results of experimental study of two-photon absorption (2PA) induced in electro-optical crystals ZnTe and ZnSe by intensive optical radiation in 450–950 nm wavelength range. These crystals are of interest due to common usage in THz systems. Femtosecond pump-probe spectroscopy with the supercontinuum source helps to induce and detect different types of non-linear processes such as 2PA and to estimate nonlinear part of refractive index.

1. Introduction

Residing in the frequency scale between IR and millimeter spectral ranges, THz radiation provides wide possibilities for studying molecular structures of materials because most vibration and rotation spectra of molecules fit this range [1]. The application of optoelectronic techniques to the generation and detection of THz radiation is now well established. Widespread use of powerful laser systems for ultrashort pulses generation [2] leads to a non-linear electro-optical response of semiconductor crystals. It is important to note that this response is important when we use samples as detectors: arising nonlinearities result in data corruption [3]. For THz emission: photo induced charge carriers screen a significant part of the generated THz radiation in the crystal thereby limiting the conversion efficiency.

Two-photon absorption is one of the main processes influencing on electro-optical sampling of semiconductor crystals. There has been observed reduction of the intensity of light propagated through the crystal and the ultra-fast response [4] of semiconductor materials. In this paper we introduce experimental results of nonlinear processes studied with femtosecond pump-probe supercontinuum spectroscopy. ZnTe [5] and ZnSe wide gap semiconductor crystals used as THz radiation sources and detectors are tested. Using supercontinuum 2PA absorption dynamics and nonlinear part of refractive index for 450–950 nm range is detected.

2. Experimental study of two-photon absorption

To investigate the nonlinear response of ZnTe and ZnSe semiconductor crystals set of experiments on the femtosecond pump-probe spectroscopy setup (Fig. 1) was held. As the result we get dependences of the absorbed radiation incident on crystals on the time delay between pulses. The main feature of

IOP Conf. Series: Journal of Physics: Conf. Series **1062** (2018) 012009 doi:10.1088/1742-6596/1062/1/012009

this setup is use of supercontinuum as a probe beam. It provides insight into induced nonlinear properties of media in 350-1400 nm range.

First results using this technique are presented in paper [6]. Pump-probe supercontinuum spectrometer was based on a Ti: Sapphire regenerative amplifier (Avesta Regulas 35F1K), which produces 30 fs pulses with energy up to 2.3 mJ and repetition rate of 1 kHz. Beam was split (*BS*1) into two parts: one was sent through the delay line and was used as a pump beam, while the second was used for supercontinuum [7] (SC) generation and after that as a probe one. Mirrors *M*1, *M*2, installed on a linear translator with stepped motor controlled by a computer, were used as an optical delay line. The maximum delay time between pump and probe pulses was 660 ps with the delay step of 6.6 fs. To produce high-intensity ultra-wide SC in the IR and UV spectral regions a water jet was applied. The supercontinuum generator consisted of *L*2 lens with 10 cm focus length and 1.8 mm width water jet placed 4 mm before the focus of *L*2 lens. *L*3 lens with 5 cm focus length collimated the resulting spectral supercontinuum. Both beams passed through a quartz lens *L*1 with a focal length of 15 cm and were combined in space on a sample located at distance of 7.5 cm from the lens. The probe beam was registered by spectrometer ASP-100M (*SP1*).

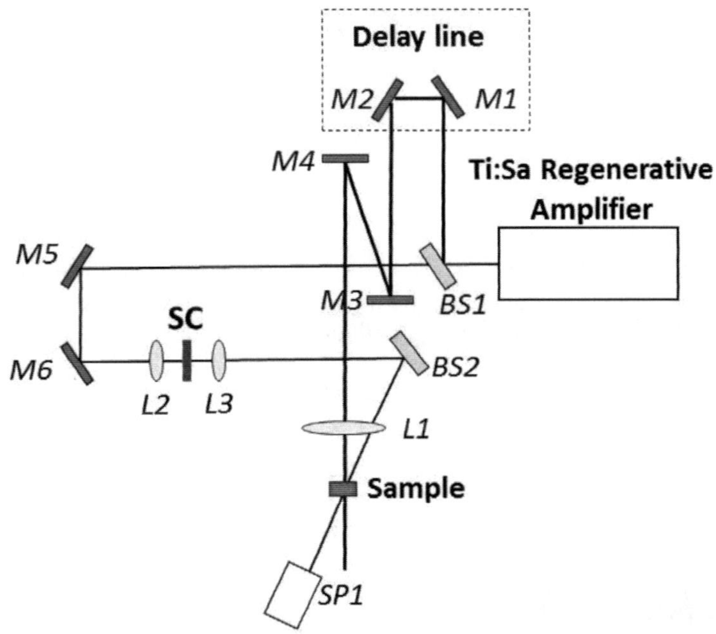

Figure 1. Pump-probe supercontinuum spectrometer setup scheme: *M1-M6* – mirrors, *L1-L3* – lenses, *BS1-BS2* – beam splitters, *SP1* - spectrometer.

By using pump-probe technique, we recorded pump-probe scans, pumping at 800 nm and probing at different frequencies and power levels of beams. In fig. 2 some results for ZnSe are presented. Dynamics of transmission of probe wavelengths 550, 650 and 900 nm for 50 mW pump pulse are indicated in the graph. As it illustrated there is a peak of 2PA and it changes it's width and position. Shift of band peak dependence on probe wavelength can be used to quantify refractive index dispersion. Band width at half-maximum wave dependence helps to determinate relaxation of induced excitation in material.

IOP Conf. Series: Journal of Physics: Conf. Series **1062** (2018) 012009 doi:10.1088/1742-6596/1062/1/012009

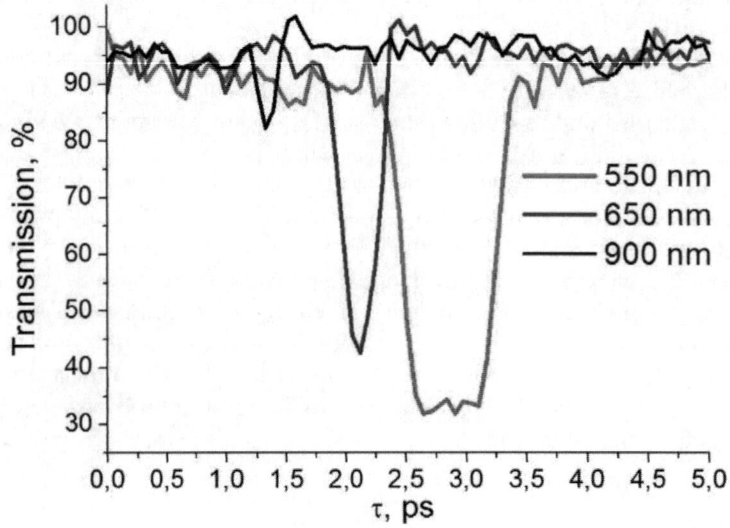

Figure 2. Pump-probe transmission dynamics for ZnSe crystal for 50 mW pump.

2.1. Characteristics of samples

As it was mentioned above, in this paper we present results of investigation of intrinsic semiconductor crystals ZnTe and ZnSe, plane plates with orientation and width as shown in Table 1. These are crystals with the typical A2B6 band structure and vibrational subsystem.

Table 1. Parameters of ZnSe and ZnTe crystals.

	Orientation	Thickness (μm)
ZnSe	[100]	560
ZnTe	[100]	710

3. Nonlinear properties of crystals

Using wavelength evolutions of 2PA absorption bands at half-maximum for ZnSe, ZnTe and GaP crystals, extracted from various power dependences, and with knowledge of sample thickness, we are able to obtain nonlinear index of refraction. Dispersion of refractive index for different pump power levels and dispersion under ordinary conditions [8] are shown on Figure 3.

As it can be seen from the refractive index dispersion graph, at the pump power level of the order of 10 GW/cm^2, the dispersion of the refractive index unalters with slight pump change. When the pump power level increases up to 45 GW/cm^2, the dispersion of the refractive index changes drastically. This can be explained by the activation of the nonlinear part of the refractive index n_2.

We can determinate the values of the nonlinear part of the refractive index n_2:

$$n_2 = \Delta n / \Delta I \quad (1)$$

In Table 2 results of estimated values of n_2 are presented as well as values obtained in earlier studies [9-11].

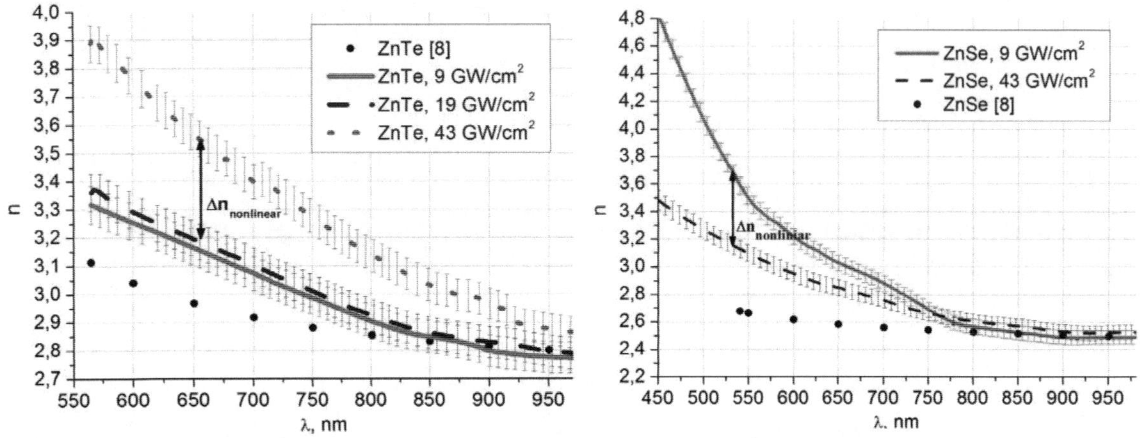

Figure 3. Dispersion of refractive index for different pump power levels.

The results are close to each other, which confirms the motivation for the use of the proposed scheme of femtosecond pump supercontinuum probe spectrometer.

Table 2. The values of the nonlinear part of the refractive index n_2.

	n_2, cm²/W					
	532 nm	532 nm, exp.	780 nm	780 nm, exp.	810 nm	810 nm, exp.
ZnSe	-4.4×10^{-11} [9]	-1.6×10^{-11}	10×10^{-12} [10]	6.5×10^{-12}	-	-
ZnTe	-	-	5×10^{-13} [11]	7.4×10^{-12}	1.5×10^{-12} [12]	7×10^{-12}

4. Conclusion

With supercontinuum in a pump-probe spectroscopy system, we investigated the behavior of two-photon resonance at different wavelengths from 450 - 950 nm in commonly usable electro-optic semiconductor crystals ZnTe and ZnSe. Varying pump power, we are able to estimate nonlinear index of refraction, which is significant in THz electro-optical sampling.

References

[1] Rostami A, Rasooli H and Baghban H 2010 *Terahertz Technology: Fundamentals and Applications* (Berlin: Springer) p 15
[2] Rulliere C 2005 *Femtosecond laser pulses* (Berlin: Springer Science)
[3] Grachev Y, Osipova M and Bespalov V 2014 *J. Quantum Electronics* **44** p 1170
[4] Hase M 2009 *Applied physics letters* **94** 112111
[5] Avetissov I, Mozhevitina E, Khomyakov A, Avetisov R, Davydov A, Chegnov V, Chegnova O and Zhavoronkov N 2015 *J. CrystEngComm* **17** p 561
[6] Osipova M, Makarov E and Bespalov V 2016 *Proc. Int. Conf. of LO (St. Petersburg)* R8-45
[7] Tcypkin A, Putilin S, Melnik M, Makarov E, Bespalov V and Kozlov S 2016 *J. Applied optics* **55** p 8390
[8] Li H 1984 *Journal of phys. and chem. ref. data* **13** p 103
[9] Said A, Sheik-Bahae M, Hagan D, Wei T, Wang J, Young J and Van Stryland E 1992 *JOSA B.* **3** p 405
[10] Krauss T and Wise F 1994 *Applied physics letters* **14** p 1739
[11] Yin M, Li H, Tang S and Ji W 2000 *Applied Physics B: Lasers and Optics* **4** p 587
[12] Xing Q, Mao F, Lang L, Wang Z, Wang K, Li S and Wang Q 2005 *Optics & Laser Technology* **6** p 475

IOP Conf. Series: Journal of Physics: Conf. Series **1062** (2018) 012010 doi:10.1088/1742-6596/1062/1/012010

Epsilon-Near-Zero metal-dielectric composite for terahertz frequency range

E A Litvinov, A V Vozianova and M K Khodzitsky

Terahertz Biomedicine Laboratory, *ITMO University*, 3b, Kadetskaya line VO, Saint-Petersburg 199034, Russia

E-mail: litviegor@yandex.ru

Abstract. Recent demonstrations of the specific class of materials such as Epsilon-Near-Zero - materials (ENZ) cause a big interest because of their potential application for manipulation and control of the propagation of the radiation on subwavelength level. The latest experiments on these materials were realized in microwave, far-infrared and visible spectral range, but there were not any demonstrations in terahertz frequency range. In this paper, we present the research of ENZ wired metal-dielectric composites with effective permittivity ranging from 0 to 1 that was designed for application as a phase compensator in terahertz frequency range. Numerical simulation of the composites was calculated by the Finite-difference time-domain method (FDTD).

Introduction

Current success in studies and development of optical metamaterials attracts special attention of the research community considering their unique properties that are not occurred in ordinary materials. Metamaterials are extremely attractive in the view of photonics. Specific effects, that are predicted in optical metamaterials permit to create not only new devices and gadgets, but to replace old ones. In particular, it is supposed to use metamaterials to manufacture both effective and exotic optical elements. One engaging type of optical metamaterials is ENZ-material («Epsilon-Near-Zero»). ENZ is the artificial material with extremely low value of the dielectric permittivity, which was studied in microwave spectral range, where ENZ properties have been predicted theoretically [1–3] and demonstrated experimentally [4–8]. ENZ materials correspond with phenomena like squeezing electromagnetic energy [3], supercoupling [9] and tunneling [2]. It can be used to obtain the images with high resolution in visible and near-infrared spectra, control of heat radiation with the purpose to avoid thermal losses or increase the efficiency of thermophotovoltaic conversion (TRV) [10]. Such new artificial media may find application in waveform shaping [1,13], obtaining tunnel effect [2], the control of spontaneous emission enhancement and superradiance [14]. This kind of metamaterial allows constructing some outstanding devices. Indeed, ENZ-media could be used for designing materials with high transmission or high reflection [11,12]. Devices based on ENZ give the opportunity to create materials that can change their properties from dielectric to metallic with the slight alteration of material parameters [15]. It was shown that ENZ can be applied in dielectric sensors [16], nanocircuits [17], Fourier transform [18] and beamshaping [19]. It was suggested to use ENZ for creating EO-modulators, nanoresonators [20] and beam splitters [21].

In recent years, terahertz (THz) technology has made great advance in the fields of biological samples analysis, THz image, high-speed wireless communication and so on. To the further development of THz technology, there is also a high demand for the tunable functional devices [22-25]. In this work we present and analyse the metal wires-dielectric composites with the effective dielectric permittivity ranged from -1 to 1 for THz frequency range.

Theory

The development of the metal-dielectric composite media was made using the effective medium theory (EMT). This theory is simple mathematically and represents us the method which allows getting the information about effective properties of the metal-dielectric composites quickly [26]. There are two approximations in EMT:

- a composite of metallic and dielectric inclusions is the homogeneous medium, which can be described from the terms of self-consistency;
- either metallic or dielectric inclusions have the same form.

In theory, composite has to be structure with period d and definite dispersion of dielectric permittivity in the frequency range 0.1-1 THz. In this work the plastic VisiJet M3 with $\varepsilon_b = 2.76$ is used for basis of the structure. With the approximation in EMT the composite based on material with permittivity ε_b with metal inclusions (wires) the dispersion can be calculated by (1):

$$\varepsilon_{eff} = \varepsilon_b - \frac{\omega_p^2}{\omega(\omega + i\gamma)}, \qquad (1)$$

where $\omega_p = \dfrac{2\pi c^2}{d^2 \ln \dfrac{2d}{d_a}}$ is the plasma frequency, $\gamma = \dfrac{4\varepsilon_0 d^2 \omega_p^2}{\pi d_a^2 \sigma}$, σ is the conductivity of metal inclusions, d_a is diameter of wires in the structure, c is the speed of light in vacuum.

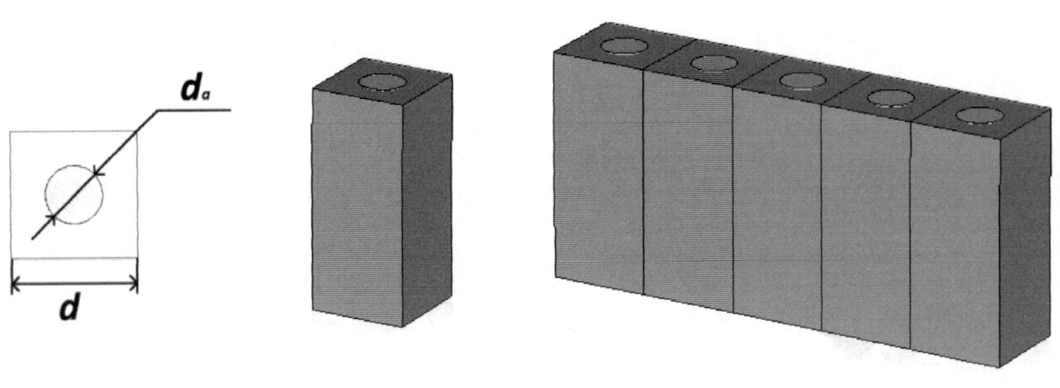

(a) (b) (c)

Figure 1. The geometry of simulated wired metal-dielectric structure. It is composed of unit cells with cylindrical copper wires with diameter $d_a = 300$ μm.

IOP Conf. Series: Journal of Physics: Conf. Series **1062** (2018) 012010 doi:10.1088/1742-6596/1062/1/012010

It was necessary to change the concentration of the metal cylinders with the same diameter d_a in different samples in order to make the composites with different effective permittivity based on plastic-photopolymer VisiJet M3. Hence, it was necessary to change the period of structure *d* (figure 1) which has the limit of manufacturing for 3D printing (30 μm).

Numerical Simulation

Simulations of the composite were calculated by the Finite-difference time-domain method (FDTD) using CST Microwave Studio. It was demanded that the electrical field of the incident wave was codirected with the orientation of wires in order to show anisotropic properties of the composite.

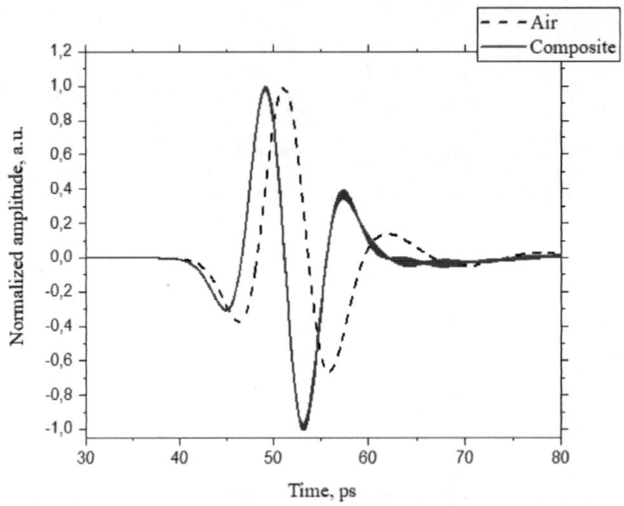

Figure 2. Pulses with bipolar profile through the air layer (black dotted line) and the composite with the period of unit cell 450 μm (blue line).

It was needed to obtain negative phase delay which provides designed material to work as phase compensator. The phase delay is meant as the 'air' pulse (propagated through the air) shifts relative to 'composite' pulse (propagated through the composite).

Table 1. Periods of composites and phase delays as a result of numerical simulations. Diameter of copper wire is 300 μm

Dimension of the unit cell *d* (μm)	Phase Delay (ps)
650	-1.203
600	-1.480
550	-1.621
500	-1.764
450	-1.768
400	-1.875

Several simulations with changing period of structure (table 1) were made both for one-layered composite and the same volume filled with the air. The phase delay was calculated using this data (table 1) by equation (2):

$$\delta t = \frac{d}{c}(n-1), \tag{2}$$

where n - the effective refractive index of a composite, c – speed of light. Dispersion curves show exactly the borders of material working as ENZ with regard to various dimensions of the unit cell. The period of structure less than 400 μm was not considered due to closeness of technological limits.

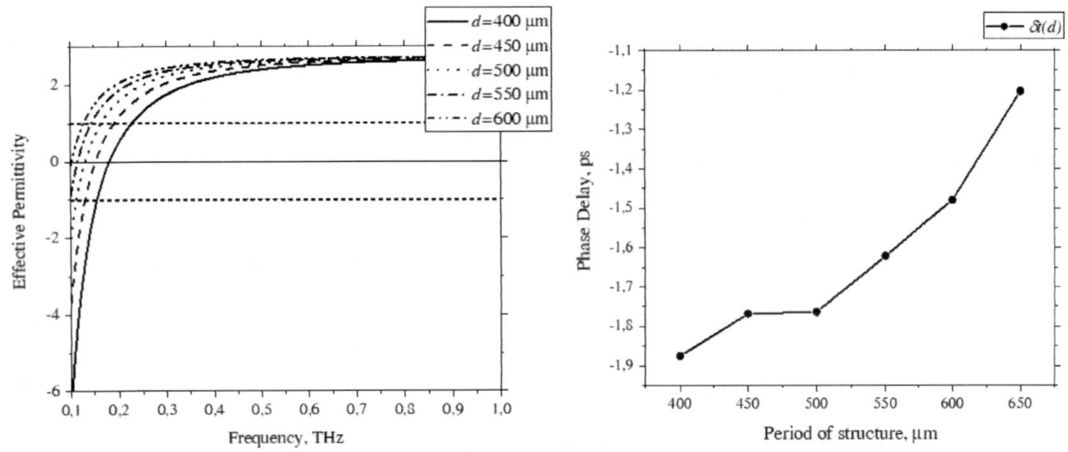

(a) (b)

Figure 3. Demonstration of ENZ-composite characteristics. Curves (a) show the dispersion of dielectric permittivity and dotted lines show the approximate borders of ENZ material. Presented results of numerical simulations (b) describe how one layer of composite can give the phase delay by decreasing the period of structure.

According to computations and simulations, phase delay of the ENZ-composite can be increased by simply reducing the period of structure.

Conclusions

In this work the calculation and numerical simulation of the ENZ metal-dielectric wired composite in terahertz frequency range were demonstrated. Simulations show the negative phase delay for each simulated sample with different periods of structure. It has been shown that the delay can be increased by simply decreasing the period of structure, but the limits of manufacturing must be taken into account. Experimental demonstration of this terahertz Epsilon-Near-Zero metamaterial is expected in the near future.

References

[1] Alù A, Silveirinha M G, Salandrino A and Engheta N 2007 *Phys. Rev. B* **75** 155410
[2] Silveirinha M G and Engheta N 2006 *Phys. Rev. Lett.* **97** 157403
[3] Alù A and Engheta N 2008 *Phys. Rev. B* **78** 035440
[4] Engheta N and Polman A 2013 *Nature Photonics* **7** 907–12
[5] Vesseur E J R, Coenen T, Caglayan H, Engheta N and Polman A 2013 *Phys. Rev. Lett.* **110** 013902

[6] Adams D C, Inampudi S, Ribaudo T, Slocum D, Vangala S, Kuhta N A, Goodhue W D, Podolskiy V A and Wasserman D 2011 *Phys. Rev. Lett.* **107** 133901

[7] Edwards B, Alù A, Silveirinha M G and Engheta N 2009 *Journal of Applied Physics* **105** 044905

[8] Khodzitsky M K, Kharchenko A A, Strashevskyi A V, Tarapov S I 2009 *Telecommunications and Radio Engineering* ***68***(7) 561-566

[9] Silveirinha M G and Engheta N 2007 *Phys Rev B* **76** 245109

[10] Dyachenko P N, Molesky S, Petrov A Y, Störmer M, Krekeler T, Lang S, Ritter M, Jacob Z and Eich M 2016 *Nature Communications* **7** 11809

[11] Xu Y and Chen H 2011 *Appl. Phys. Lett.* **98** 113501

[12] Luo J, Xu P, Chen H, Hou B, Gao L and Lai Y 2012 *Appl. Phys. Lett.* **100** 221903

[13] Luo J, Xu P and Gao L 2012 *J. Opt. Soc. Am. B* **29** 35–39

[14] Alù A and Engheta N 2009 *Phys. Rev. Lett.* **103** 043902

[15] Kaipurath R M, Pietrzyk M, Caspani L, Roger T, Clerici M, Rizza C, Ciattoni A, Di Falco A and Faccio D 2016 *Scientific Reports* **6**

[16] Alù A and Engheta N 2008 *Phys Rev B* **78** 045102

[17] Engheta N 2007 *Science* **317** pp 1698–1702

[18] Navarro-Cía M, Beruete M, Sorolla M and Engheta N 2012 *Phys. Rev. B* **86** 165130

[19] Torres V, Pacheco-Peña V, Rodríguez-Ulibarri P, Navarro-Cía M, Beruete M, Sorolla M and Engheta N 2013 *Opt. Express* **21** 9156–9166

[20] Yao J, Yang X, Yin X, Bartal G and Zhang X 2011 *Proceedings of the National Academy of Sciences* **108** 11327–31

[21] Gurvitz E A, Vozianova A V, Khodzitsky M K 2014 *Journal of Physics: Conference Series* ***541***(1) 012067

[22] Denisultanov A Kh, Khodzitsky M K 2013 *Proceedings of SPIE - The International Society for Optical Engineering* **8806** 88062I

[23] Girich A, Khodzitsky M, Nedukh S, Tarapov S 2011 *NATO Science for Peace and Security Series B: Physics and Biophysics* 159-164

[24] Tarapov S I, Khodzitsky M, Chernovtsev S V, Belozorov D, Merzlikin A M, Dorofeenko A V, Vinogradov A P, Inoue M, Granovsky A B 2010 *Physics of the Solid State* ***52***(7) 1427-1431

[25] Tarapov S I, Khodzitskiy M, Chernovtsev S V, Belosorov D, Merzlikin A M, Vinogradov A P, Granovsky A B 2009 Solid State Phenomena ***152-153*** 394-396

[26] Pendry J B, Holden A J, Robbins D J and Stewart W J 1998 *Journal of Physics: Condensed Matter* **10** 4785

Development methodology definition diameter of particles in the composite medium using pulsed terahertz spectroscopy

A V Salnikova[1], M Shui[2], K I Timoshchuk[1] and M K Khodzitsky[1]

[1]ITMO University, Saint-Petersburg, Russia
[2]University of Rochester, Rochester, USA.

alina.v.salnikova@gmail.com

Abstract. In this paper, a method for determination of the diameter of spherical particles in composite medium on the basis of the Mie scattering theory using terahertz spectroscopy in the range from 0.1 THz to 1 THz was proposed.

1. Introduction

Terahertz (THz) light is a million times lower frequency than X rays and is non-ionizing. Therefore, it would be safe and desirable to use THz light for medical therapy and diagnostic purposes [1-4] for which it needs to develop new THz components [5-7]. Spectroscopic measurements using terahertz radiation are highlighted as they probe the scattering properties of granular media, which are sensitive to the packing structure. The wavelength of terahertz emission ranges from 30 μm to 3 mm and correlates with the particle size in many typical granular (composite) media. Many parameters obtained with angular scattering and spectral attenuation or in the measurement of reflection become sensitive to particle size and to other geometric characteristics of the samples, such as the degree of polarization or the intensity of scattering and damping. There is possibility of estimating the characteristics of composite media using THz Time-Domain spectroscopy (TDS). The applying of TD spectroscopy is not limited to the study of the quality of manufacturing of artificial media: it is used to analyzing both biological media and food products. A large number of studies make it possible to estimate the accuracy of the obtained results and the possibility of their application for improving quality control or monitoring of manufacturing process [8]. In this paper, the THz technique for analyzing composite media which contain spherical particles of a certain radius was proposed. The dependence of the scattering and absorption cross section on the radius of spherical particles for metals, semiconductors, and dielectrics was investigated based on the theory of Mie.

2. Theory

The analytical method is based on the Mie Theory. The Mie theory of the scattering of radiation on sphere in a homogeneous medium is based on the expansion of the electromagnetic field along the cylindrical harmonics and the "cross-linking" of the tangential components of the electric and magnetic field strengths at the boundary of the sphere. To satisfy these boundary conditions, it is necessary to assume that, in addition to the field incident on the sphere and the field inside the sphere, there is also the field of the secondary (scattered) wave [9]. Following formula (1) is used to calculate the cross section scattering of radiation for metal sphere according the Mie theory

$$\sigma_{scatt} = \frac{2\pi c^2}{\omega^2 \varepsilon_m} \sum_{n=1}^{\infty} (2n + 1)\{|a_n(x, mx, m)|^2 + |b_n(x, mx, m)|^2\}, \qquad (1)$$

where c is the speed of light in vacuum, ε_m is the permittivity of the medium, ω is the cyclic frequency of the incident radiation, a_n and b_n are the Mie coefficients. The formula (2) is used to calculate the coefficient x

$$x = kr_s, \qquad (2)$$

where r_s is the radius of the sphere, k is the wave vector of radiation. Next formula (3) is used to calculate the coefficient m:

$$m = \sqrt{\frac{\varepsilon_s(\omega)}{\varepsilon_m}}, \qquad (3)$$

where ε_s is the permittivity of the material of the sphere, ε_m is the permittivity of the matrix. The formula (4) and (5) used to calculate the Mie coefficients:

$$a_n(x, y, m) = \frac{\psi_n'(y)\, \psi_n(x) - m\, \psi_n'(x)\, \psi_n(y)}{\psi_n'(y)\zeta_n(x) - m\, \zeta_n'(x)\psi_n(y)}, \qquad (4)$$

$$b_n(x, y, m) = \frac{m\, \psi_n'(y)\psi_n(x) - \psi_n'(x)\, \psi_n(y)}{m\psi_n'(y)\zeta_n(x) - \zeta_n'(x)\psi_n(y)}, \qquad (5)$$

where $\psi_n(z)$ and $\zeta_n(z)$ are the Debye functions. The formula (6) and (7) are used to calculate the Debye functions:

$$\psi_n(z) = zj_n(z) = \sqrt{\frac{\pi z}{2}} J_{n+\frac{1}{2(z)}}, \qquad (6)$$

$$\zeta_n(z) = zh_n^{(1)}(z) = \sqrt{\frac{\pi z}{2}} H^{(1)}_{n+\frac{1}{2(z)}}, \qquad (7)$$

where $j_n(z)$ is the spherical Bessel function, $J_{n+1/2(z)}$ and $H^{(1)}_{n+1/2(z)}$ are the functions of Bessel and Hankel for half-integral order. The formula (8) is used to calculate the cross-section of extinction. Extinction describes the decrease of the intensity of radiation as a result of absorption and scattering on particles.

$$\sigma_{ext} = \frac{2\pi c^2}{\omega^2 \varepsilon_m} \sum_{n=1}^{\infty} (2n + 1) Re\{a_n(x, mx, m) + b_n(x, mx, m)\}. \qquad (8)$$

The formula (9) is used to calculate the cross section of absorption. The radiation absorption cross section is obtained from the difference between the extinction and scattering cross sections.

$$\sigma_{abs} = \sigma_{ext} - \sigma_{scatt}. \qquad (9)$$

This theory can help analytically study the dependence position of the scattering peak and the radius of a spherical particle.

3. Materials and methods

In the experiment the setup, which scheme is shown in Figure 1, was used. Broadband pulsed THz radiation was generated using an undoped InAs crystal by irradiating it with femtosecond pulses of an ytterbium laser (wavelength - 1040 nm, pulse duration - 120 fs, pulse repetition frequency - 75MHz, power - 1 W). THz radiation had the following output characteristics: spectral range from 0.05 to 2 THz, average power - 30 uW, pulse duration - 2.7 ps. The main power was concentrated in the frequency range from 0.12 to 1.1 THz. THz radiation was generated by a gallium arsenide crystal in the magnetic system, then passed through a Teflon filter (which cuts wavelengths shorter than 50 um). After that, the radiation passed through the sample fixed in a focal plane perpendicularly to the beam. With the simultaneous incidence of the femtosecond probe beam and THz beam passed through the sample on the electro-optical CdTe crystal, THz pulse induced birefringence of the probe beam in the crystal due to the electro optical effect. The birefringence magnitude is directly proportional to the intensity of terahertz wave electric field in the time point E (t) [10].

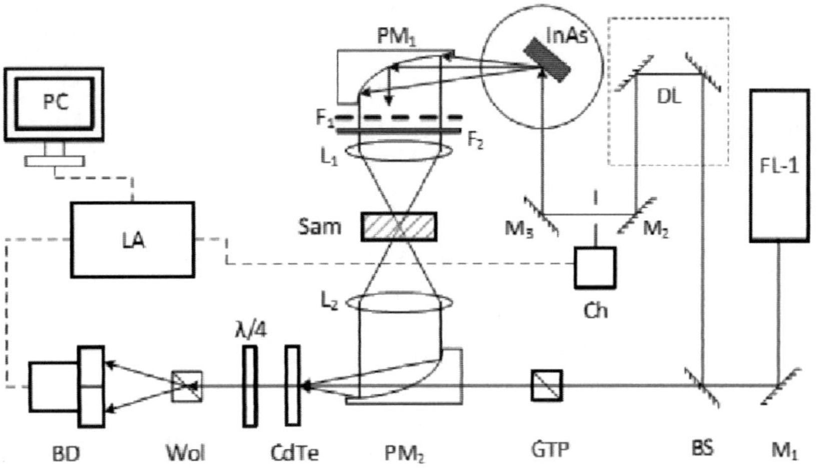

Figure 1. Schematic diagram of the set up (FL-1 - femtosecond laser based on potassium-yttrium tungstate crystal activated with ytterbium (Yb: KYW), generating femtosecond pulses; $F_{1,2}$ – a set of teflon filters for IR wavelength range cutting off, BS – beamsplitter, DL – optical delay line, $M_{1,2,3}$ – mirrors, Sam – investigated sample, Wol – Wollaston prism, CdTe – electro optical cadmium-telluric crystal, BD – balanced detector, LA – lock-in amplifier, PC – personal computer, GTP - Glan-Taylor prism, $PM_{1,2}$ – parabolic mirrors, Ch - chopper.

The experiment structure was shown in Figures 2-3. This structure was made of VisiJetM3 Crystal. There are hemispherical depressions of various diameters: 0.2 mm, 0.4 mm, 0.6 mm, 0.8 mm, 1 mm and one section of the structure without depressions.

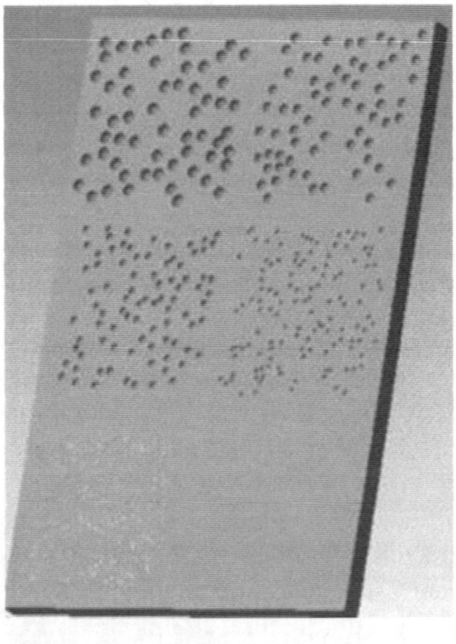

Figure 2. Image of model the experimental structure

Figure 3. Photo of the experimental structure

During the experiment, the structure was arranged in such a way that the radiation fell on the region in which the spheres of the same diameter were. Knowing the diameters of spherical particles in each region, it is possible to experimentally verify the size of spherical particles in a given composite medium. Since the structure itself should be an analog of the composite medium, the two printed plates of the experimental structure were compressed in such a way that the spherical particles of air were inside the structure.

4. Results and analysis

Using the Mie theory the dependences of the resonance frequency on the radius of air spherical particles for 3 kinds of plastic (TPX, PLA, VisiJetM3 Crystal) were calculated. After carrying out analytical studies the graph of the dependence of the scattering peak frequency on the radius of spherical particles in the experimental medium (Figure 4) was plotted, the obtained graph can be approximated into an exponential dependence. According to the graph, one can see that as the particle diameter increases, the resonant frequency of the scattering peak shifts to the low-frequency range. As a result, the radius of the particles was determined from the experimentally obtained frequency of the scattering peak, using the available analytical calculations.

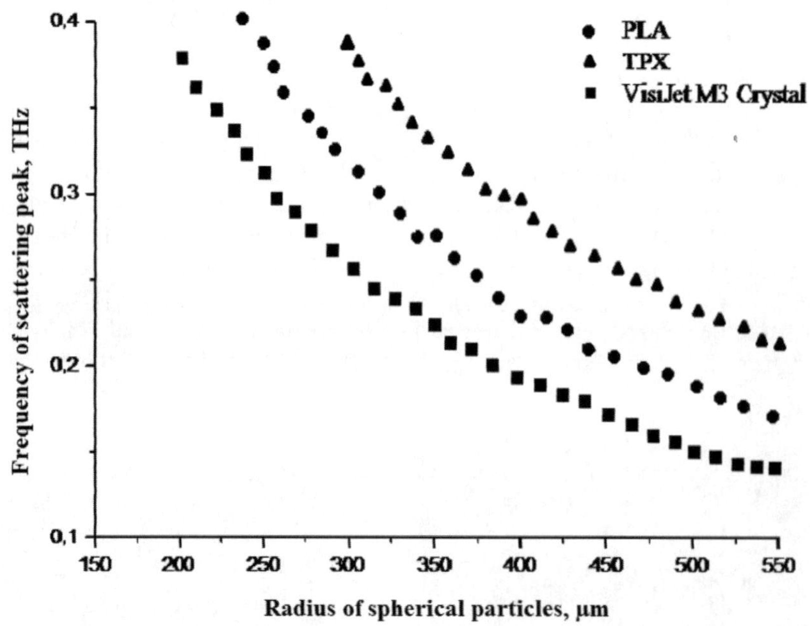

Figure 4. Analytical dependence of the scattering peak frequency on the radius of air particles in the matrices

When analyzing the reflection spectrum of the composite medium, the reflection peak was found to be at a frequency corresponding to particles of smaller radius. With the help of the conducted analytical modeling it was revealed that the peak for the sector with spherical particles with a diameter of 1 mm is at a frequency corresponding to particles with a diameter of 0.8 mm. Using a microscope, the diameter of the spherical particles was measured, and it turned out that their diameter was not 1 mm, but 0.8 mm (Figure 5). Thereby, it can be argued that this technique can be used to determine the diameter of particles in composite media by means of pulsed terahertz spectroscopy. As a result of the experiment and the analysis of the results obtained using the developed methodology, it was found that the fabricated structure did not correspond to the simulated structure: the particle radii were less than expected.

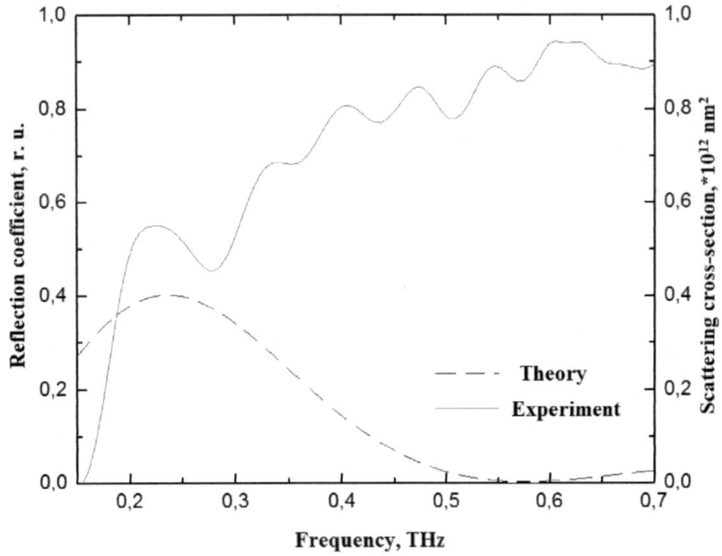

Figure 5. The reflection spectrum of experimental structure for particles with diameter of 1mm and the scattering cross-section calculated using the Mie theory for particles with diameter of 0.8 mm

5. Conclusion

The Mie theory makes possible to estimate the diameter of spherical particles in matrix using the frequency position of the scattering and absorption peak. In this paper, the diameter of particles in dielectric medium was measured using THz time-domain spectroscopy based on Mie theory.

References

[1] Gusev S I, Borovkova M A, Strepitov M A, Khodzitsky M K 2015 *Progress in Biomedical Optics and Imaging - Proceedings of SPIE*, **9537**, 95372A

[2] Duka M V, Dvoretskaya L N, Balbekin N S, Khodzitskii M K, Chivilikhin S A, Smolyanskaya O A 2014 *Quantum Electronics*, **44**(**8**), 707-712

[3] Borovkova M, Serebriakova M, Fedorov V, Sedykh E, Vaks V, Lichutin A, Salnikova A, Khodzitsky M 2017 *Biomedical Optics Express*, **8**(1), 273-280

[4] Borovkova M, Khodzitsky M, Demchenko P, Cherkasova O, Popov A, Meglinski I 2018 *Biomedical Optics Express*, **9**(5), 2266-2276

[5] Denisultanov A Kh, Khodzitsky M K 2013 *Proceedings of SPIE - The International Society for Optical Engineering*, **8806**, 880621.

[6] Khodzitsky M K, Vozianova A V, Gill V V, Chernyadiev A V, Grebenchukov A N, Minin I V, Minin O V 2016 *Proceedings of SPIE - The International Society for Optical Engineering*, **9918**, 99182X.

[7] Gill V V, Vozianova A V, Khodzitsky M K 2015 *Journal of Physics: Conference Series*, **643**(1), 012052

[8] Bandyopadhyay A et. al. 2007 *International Journal of Infrared and Millimeter Waves* **11** 969–978

[9] Hulst Van De H C 1957 Light scattering by small particles **12** pp 28–30

[10] Bespalov V G, Gorodetski A A and Denisyuk I Yu 2008 *J. Opt. Technol.* **75**, 636–642

IOP Conf. Series: Journal of Physics: Conf. Series **1062** (2018) 012012 doi:10.1088/1742-6596/1062/1/012012

Calculation of chiral metasurface characteristics in terahertz frequency range

M S Masyukov[1], A V Vozianova[1], A N Grebenchukov[1] and M K Khodzitsky[1]

[1]Terahertz Biomedicine Laboratory, ITMO University, 49 Kronverksky Pr., St. Petersburg, 197101, Russia

E-mail: msmasyukov@corp.ifmo.ru

Abstract. In this work the ellipticity and quantity of optical activity for chiral metasurface in terahertz frequency range were obtained. The analysis of the changing of metasurface polarizing properties depending on gammadion petal ellipse was made. The numerical simulation for structure under investigation using finite element method was performed.

1. Introduction

Although chirality has been known for a long time, recently chiral structures are of interest both for fundamental physics and for applications [1-2]. Chirality is a property of asymmetry when an object is distinguishable from its mirror image. Hence, breaking the symmetry enables the manipulation of polarization states. Terahertz (THz) technology has made great advance in the fields of biological samples analysis, THz imaging, high-speed wireless communication and so on. To the further development of THz technology, there is also a high demand for the tunable and functional devices [3-7]. This is especially interesting for control and manipulation of the polarization conversion at THz frequencies [8-10] because of their significant applications [11-14]. One of the chiral structure designs under investigation is an array of twisted gammadion crosses with fourfold rotation symmetry. It was shown strong rotator power in the visible and near-IR spectral ranges for such double –periodic bilayered structure [15]. It was experimentally and numerically demonstrated the possibility of a chirality induced by negative refractive index [16]. The influence of twisted gammadion crosses curvature on polarization rotation was shown in [17-18]. It was shown [19] the ability of polarization properties control by gammadion petal width changing.

In this work we consider chiral metasurface consisted of periodic array of twisted gammadion crosses placed on silicon substrate at THz frequencies. We studied the influence of the ratio of the ellipses semiaxes forming the petal of the gammadion on metasurface chiral properties at the operating frequencies range of 0.05-0.35 THz.

2. Description of the structures

The unit cell of the structure under investigation is four double truncated ellipses forming four petals of the gammadion, as shown in figure 1. The gammadion petals are made of metallic (PEC) twisted strips width of 7 μm which are located on a silicone (Si) substrate with permittivity of 11.56. The substrate thickness is h=150 μm, a side of the unit cell is d=600 μm. The chiral metasurface was designed to operate in the frequency range of 0.1-1 THz.

The ellipse semiaxis *b* along the y-axis has a fixed length of *150 μm*, and the ellipse semiaxis *a* along the x-axis changes, see figure 1 (b)-(d).

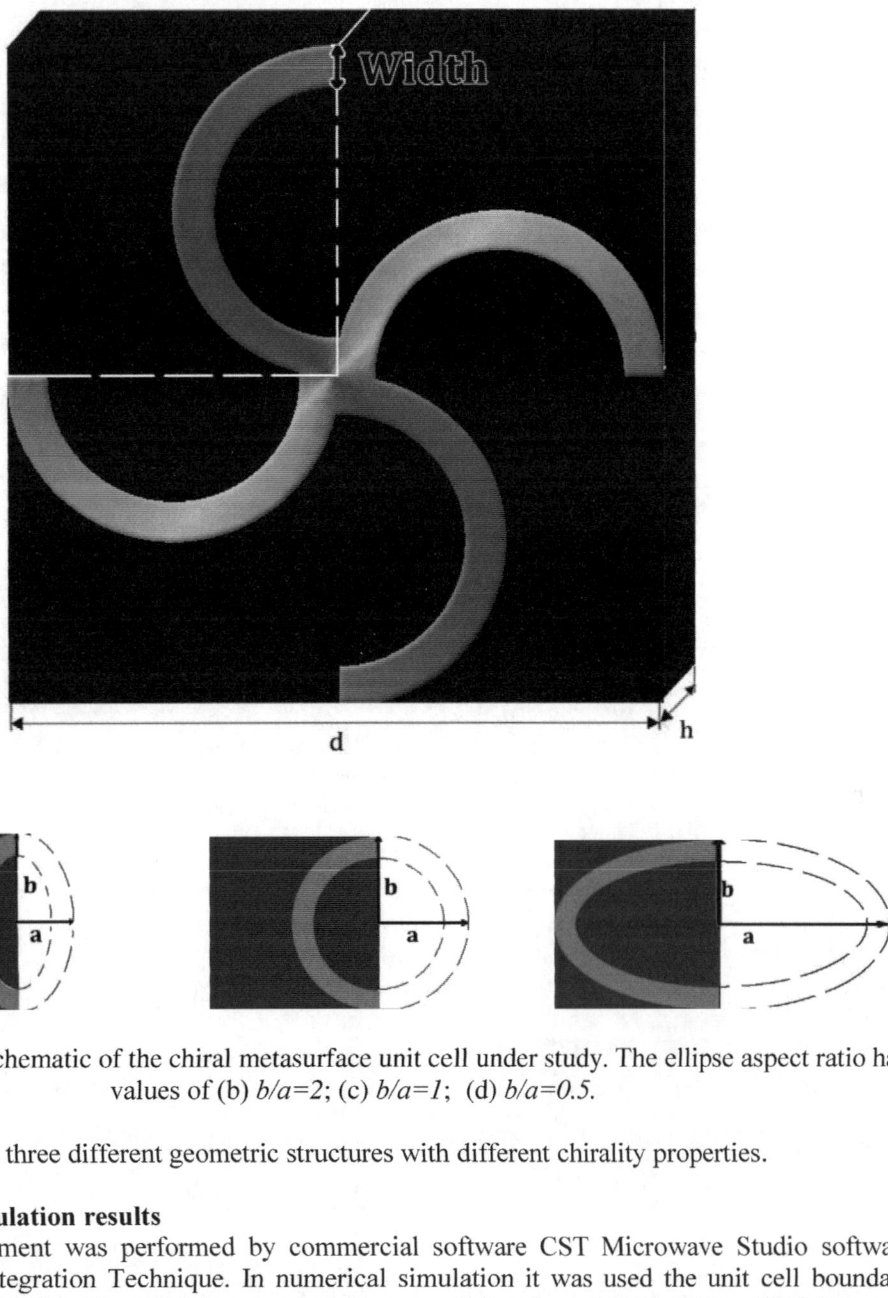

Figure 1. (a) Schematic of the chiral metasurface unit cell under study. The ellipse aspect ratio has values of (b) *b/a=2*; (c) *b/a=1*; (d) *b/a=0.5.*

Thus, we obtain three different geometric structures with different chirality properties.

3. Numerical simulation results

The virtual experiment was performed by commercial software CST Microwave Studio software based on Finite Integration Technique. In numerical simulation it was used the unit cell boundary conditions in *x* and *y* directions and the open boundary conditions in *z* direction. The ellipticity angle and the azimuth polarization rotation angle are derived using Jones matrix [20] in terms of transmission of the left- and right-handed circularly polarized waves (T++,T—) in the frequency range of 0.05 – 0.35 THz. The peculiar spectra of ellipticity angle for the different ratios of the ellipses semiaxes forming the the gammadion petal on the metasurface are shown in figure 2.

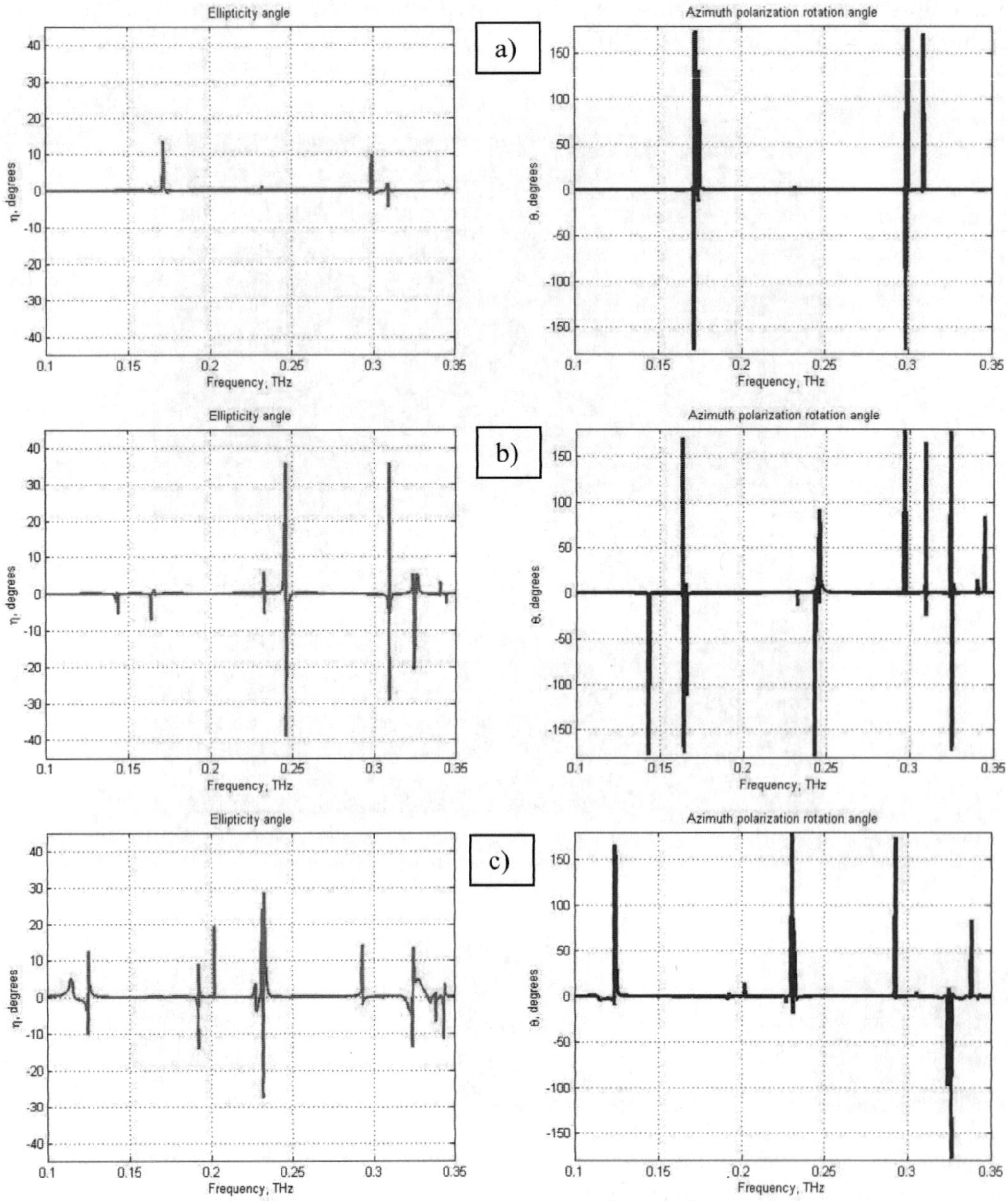

Figure 2. The ellipticity angle of metasurface for the ellipse aspect ratio forming the petal of the gammadion: a) *b/a>1*; b) *b/a=1*; c) *b/a<1*.

As seen from figure 2, there are several especial values of frequency in vicinity of which the sharp changing of polarization properties is observed. The changing of the ellipse aspect ratio (hereinafter referred to as "the ratio") leads to shifting of the values of these frequencies and changing of the metasurface polarization properties. In figure 2-a) one can see ellipticity angle and azimuth polarization rotation angle for the metasurface with the ratio of b/a=2. In this case ellipticity angle values are low (maximal $|\eta|=14$ degrees), but the structure shows giant optical activity at three resonant frequencies. Two of these extremes have positive maxima and negative minima. Such metasurface might be used as vector **E** rotator. The opposite ratio (b/a=0.5) shows a different situation (figure 2-c). The number of resonant frequencies increased, as well as the values of angle of ellipticity (maximal |η|=39 degrees). In this case azimuth polarization rotation angle has only positive or negative extremes at the same frequency, unlike the previous metasurface. This design can be used in development of multiband polarizer. The amount of ellipticity angle resonant frequencies in figure 2-b) is between the first and the second cases described above. It can be seen that the ellipticity angle values of the metasurface with circular gammadion are bigger than the ones with elliptical gammadion. The graph with azimuth polarization rotation angle looks like the superposition of such graphs in figure 2-a) and 2-c). So, this metasurface can work not only as elliptical polarizer, but as multiband linear polarization rotator.

4. Conclusions

In this work the influence of the ellipse aspect ratio forming the petal of the gammadion on metasurface chiral properties at the operating frequencies range of 0.05-0.35 THz was studied. Hence, in this paper we demonstrate polarizing properties changes caused by changes of the ellipse aspect ratio. Therefore it is possible to predict the required polarizing properties of metasurface and make an elliptic polarizer or linear polarization rotator, or make one metasurface which can be used as both of them.

Acknowledgements

This work was supported by Government of Russian Federation (Grant 074–U01).

References

[1] Wang Z, Cheng F, Winsor T and Liu Y 2016 *Nanotechnology* **27** 412001
[2] Chen H T, Taylor A J and Yu N 2016 *Reports on Progress in Physics* ***79*** 076401
[3] Terekhov Y E, Khodzitsky M K, Grachev Y V, Sedykh E A, Belokopytov G V, Zhang X-C 2013 *Proceedings of SPIE - The International Society for Optical Engineering* ***8806*** 88062Q
[4] Denisultanov A Kh, Khodzitsky M K 2013 *Proceedings of SPIE - The International Society for Optical Engineering* **8806** 880621
[5] Girich A, Khodzitsky M, Nedukh S, Tarapov S 2011 *NATO Science for Peace and Security Series B: Physics and Biophysics* 159-164
[6] Tarapov S I, Khodzitsky M, Chernovtsev S V, Belozorov D, Merzlikin A M, Dorofeenko A V, Vinogradov A P, Inoue M, Granovsky A B 2010 *Physics of the Solid State* ***52***(7) 1427-1431
[7] Tarapov S I, Khodzitskiy M, Chernovtsev S V, Belosorov D, Merzlikin A M, Vinogradov A P, Granovsky A B 2009 Solid State Phenomena ***152-153*** 394-396
[8] Cong L et. al. 2014 *Laser & Photonics Reviews 8* 626-32
[9] Huang Y et. al. 2017 *Carbon* ***119*** 305-13
[10] Xu W et. al. 2017 *Advanced Optical Materials* **5**
[11] Borovkova M. et. al. 2018 *Biomedical Optics Express* **9** pp 17039-51
[12] S. I. Gusev et. al. 2018 *Chinese Optics*, **11** 182-289
[13] Debus C and Bolivar P H 2007 *Applied Physics Letters* **91**, 184102
[14] Piesiewicz R et. al. 2007 *IEEE Antennas and Propagation Magazine* **49** pp 24-39
[15] Plum E, Fedotov V A, Schwanecke A S, Zheludev N I and Chen Y 2007 *Applied Physics Letters* **90** 223113
[16] Plum E et. al. 2009 *Physical Review B* ***79*** 035407
[17] Korolenko S Y, Grebenchukov A N, Masyukov M S, Vozianova A V and Khodzitsky M K 2016 *J. Phys.: Conf. Series* **735** 012089
[18] Masyukov M S, Vozianova A V, Grebenchukov A N and Khodzitsky M K 2017 *Progress In Electromagnetics Research Symposium-Spring (PIERS)* 3226-28
[19] Masyukov M S, Vozianova A V, Grebenchukov A N and Khodzitsky M K 2017 *Proceedings of SPIE - The International Society for Optical Engineering*, 10343, 1034338
[20] Menzel C, Rockstuhl C and Lederer F 2010 *Physical Review A* ***82*** 053811

IOP Conf. Series: Journal of Physics: Conf. Series **1062** (2018) 012013 doi:10.1088/1742-6596/1062/1/012013

Influence of the incidence radiation polarization on the absorptivity of Electrical Ring Resonator Metasurface in Terahertz frequency range.

D Gomon, E Sedykh, V Soboleva, A Zaitsev and M Khodzitsky

ITMO University, Saint Petersburg, 197101, Russia

E-mail: GomonDA89@ya.ru

Abstract. The aim of this research is investigation of the incidence radiation polarization influence on electrical ring resonators metasurface absorption in terahertz frequency range. Reflection spectra of 5 metasurface samples in cases of 3 different polarization angles ($\theta = 0^{o}$, $\theta = 45^{o}$ and $\theta = 90^{o}$) were experimentally investigated. It was shown that maximal absorption reaches in cases of polarization angles θ=0° and θ=90°.

1. Introduction

Metamaterial devices for terahertz (THz) frequency band have been a subject of close attention for the last decade. A reason behind this interest is a lack of natural materials with strong response to electromagnetic radiation of terahertz frequencies. Particularly there is a need for materials with strong absorption in terahertz frequency range for such applications as terahertz sensing, hyperspectral imaging and screening [1-7] and biomedicine applications [8-11]. The pioneering paper on THz metamaterial absorbers (MA) based on electrical ring resonators (ERR) has been published by Tao [12] (his colleague Landy studied the same structure for far infrared wavelength region [13]). This design can have theoretical absorptivity close to unity (0.99) with full width at half maximum (FWHM) of 4 percent of resonant frequency. The authors verified this design by obtaining experimentally the absorptivity value of 0.7 at 1.4 THz. Such great difference between theoretical and experimental results is explained by variation of material parameters from tabular values and by non-optimized fabrication process. In 2009 Landy [14] proposed another structure, an array of unit cells, which were comprised of two different cross-shaped resonators separated by a dielectric spacer. This MA had experimental absorptivity of 0.75 at 1.15 THz. This structure is polarization-insensitive due to the symmetry of the unit cell. In 2011 Shen et al. [15] and Ma et.al [16] proposed similar MA designs with unit cell comprised of concentric square-ring resonators backed by a layer of metal and spaced by a dielectric layer. This structure acts like a multiband absorber having three absorption peaks equal to the number of square ring resonators. This design is polarization-insensitive and gives absorptivity values of 0.99, 0.96, and 0.98 at 4.02GHz, 6.75GHz and 9.24GHz respectively. Another interesting idea to implement a dual-band absorber was proposed by Wang et al. [17]: the dual-band MA (metasurface-dielectric-metallic ground plane structure) with cut-wire metasurface unit cell. The high-order resonance of the cut-wire was used to obtain the second absorption band of high quality (Q=8.4). The simulation results give the absorptivity of 0.9967 and 0.9973 at 1.42 THz and 2.99 THz with FWHM of 0.2 THz and 0.05 THz respectively.

In this paper, the effects of the incidence radiation polarization on MMA's absorptivity were described and analyzed.

2. Description of the structure

The proposed structure consists of four layers. Layers functions, materials and thicknesses are presented in table 1.

Table 1. Layers functions, materials and thicknesses.

Layer	Function	Material	Thickness, μm
1	Substrate	Silicon	300
2	Thin film (THz mirror)	Copper	0.5
3	Spacer (absorber)	Polyimide SU-8	20
4	ERR (resonator)	Copper	0.5

The draft of the MA unit cell with double ERR is shown in figure 1. It consists of 2 ERRs connected to each other of one side and could be defined by a set of parameters: g – an ERR gap; w – an ERR lines width; a, b – ERR length and width respectively; p – a distance between an ERR edge and an unit cell border.

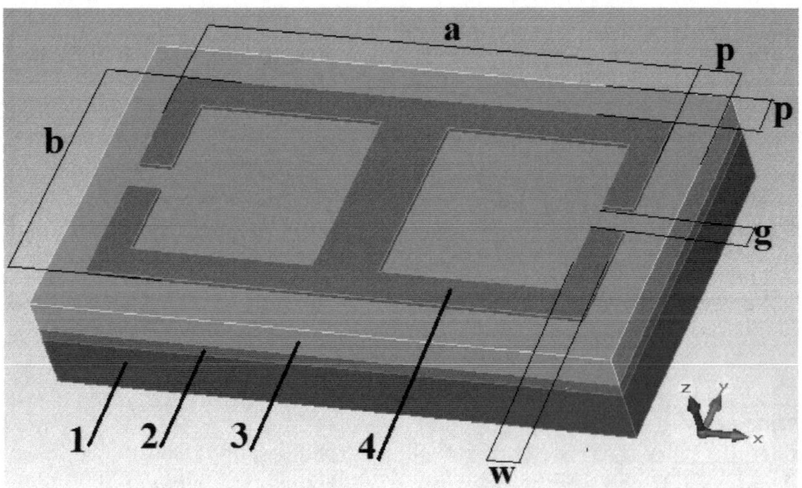

Figure 1. 3D view of MA unit cell model.

Moreover for description of incidence radiance polarization the parameter θ – an angle between short side of unit cell (Y-axis in figure 1) and electric field vector $\vec{E}$ was added.

3. Fabrication Process

For fabrication of MA, it needs the following metal-semiconductor structure: 5 copper ERR arrays (12x12 mm^2) with different geometric parameters from table 2 would be marked in silicic 320 mm substrate (because it's material with the best cost/conductivity ratio), using vacuum evaporation and photolithography process. Photolithography is profitable in case of development of tunable MA: besides high accuracy of fabrication, photolithography allows manufacturing several absorbers on one substrate simultaneously using common emulsive photomask, that increases rate of manufacture and allows cheaply making multitude of experimental samples. At the same time photolithography processes (polyimide laying, exposure, development and etching) occupies plenty of time. For the time being emulsive photomask and experimental sample of tunable MA were manufactured in"Vavilov State Optical Institute". Process of MA sample fabrication schematically is sketched in figure 2.

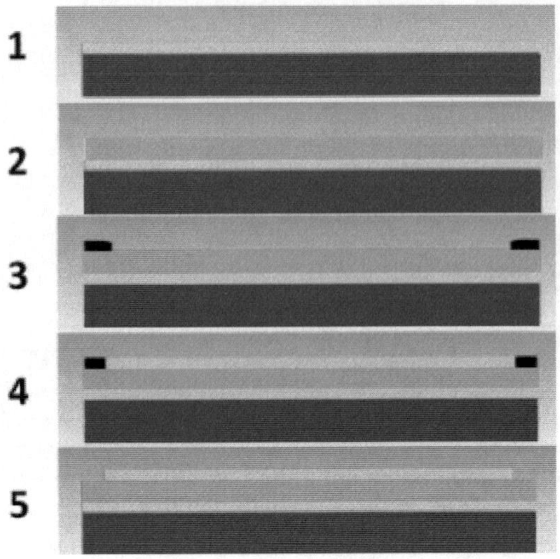

Figure 2. Process of experimental samples manufacturing consists of the following:
- substrate preparing: firstly thin chrome layer (20 nm) and after that copper layer (500 nm) was sputtered on silicon substrate, using ion-vacuum spattering method;
- liquid polyimide solution SU-8 was sputtered using centrifuge;
- photoresist layer was applied by centrifuge and image was formed by photolithography;
- another copper layer was sputtered (0.5 mm) as electronic resonant rings;
- photoresist deleting.

5 MMA samples with different geometric parameters were fabricated for investigation of influence of the incidence radiation polarization on the absorptivity of ERR Metasurface in THz frequency range. Their layer thicknesses are shown in table 1, ERR geometric parameters are presented in table 2.

Table 2. ERR geometric parameters.

MMA sample	a, μm	b, μm	g, μm	p, μm	w, μm
1	160	80	15	15	5
2	176	88	15	15	5.5
3	192	96	15	15	6
4	208	104	15	15	6.5
5	224	112	15	15	7

4. Experimental setup

In the present work compact TPS (terahertz pulsed spectroscopy) setup was used [18]. The setup operates in reflection mode and allows measuring the THz spectra between 0.25 and 1.75 THz. Figure 3 represents the experimental setup design and illustrates the process of TPS waveform detection. The TPS system makes use of the second harmonic radiation of the Er-doped femtosecond fiber laser with an average power of 200 mW for both THz pulse generation and detection. The central wavelength is 780 nm, and the pulse duration and the pulsed repetition rate are 100 fs and 50 MHz correspondingly. THz-wave generation is produced in a LT-GaAs photoconductive antenna (PCA), and detection of THz pulses is performed in a ZnSe-electro-optical detector. The reflection mode measurements with a beamsplitter and a normal angle of radiation incidence on the sample surface were carried out. Figure 3 shows the details of the experiment. The power reflectivity of the metasurface was determined by the Eq. (1):

$$R(\nu) = \left|\frac{\tilde{E}_S(\nu)}{\tilde{E}_R(\nu)}\right| \quad (1)$$

where E_R is derived from fourier-domain representation of the reference waveform $E_R = E_R(t)$, reflected from the plane gold mirror places behind the rigidly fixed diaphragm, and E_s is derived from

IOP Conf. Series: Journal of Physics: Conf. Series **1062** (2018) 012013 doi:10.1088/1742-6596/1062/1/012013

the fourier-domain representation of the sample waveform $E_S=E_S\ (t)$, reflected from the sample of interest placed behind of diaphragm.

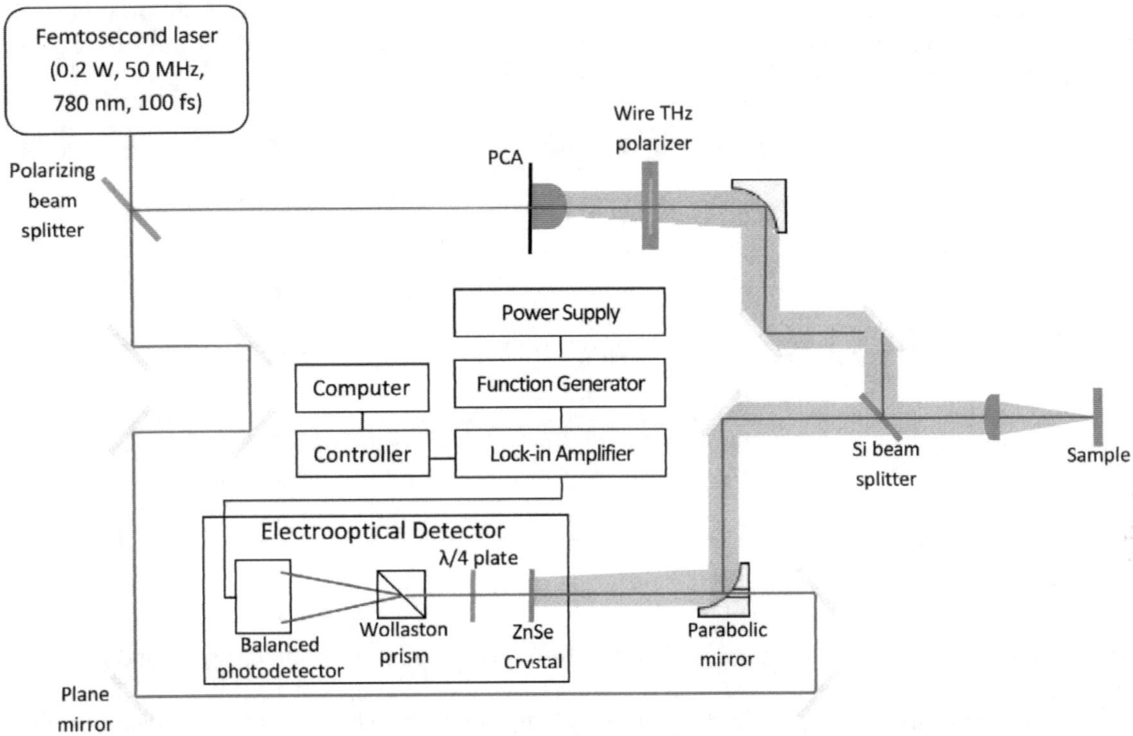

Figure 3. Design of reflective THz spectrometer.

5. Experimental results

Reflection spectra of 5 fabricated samples (geometric parameters indicated in table 2) were investigated in the frequency range of 0.1-1.6 THz for three polarization angles: θ=0°, θ=45° and θ=90°. Figure 4 demonstrates spectra of MA samples №1 and №3.

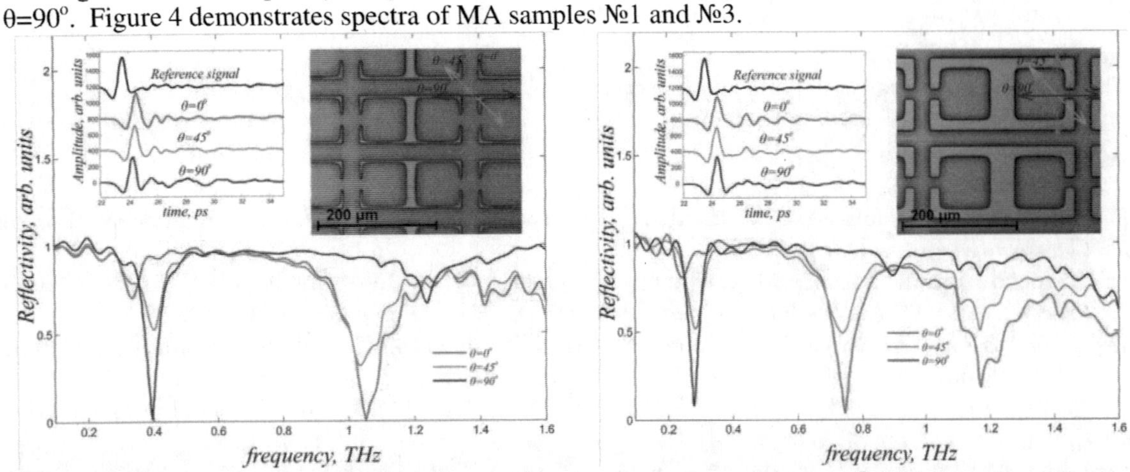

Figure 4. The reflection coefficient of the experimental MMA samples №1 and №3 for θ = 0° (red curve), θ = 45° (green curve) and θ = 90° (blue curve).

All structures spectra have 2 resonant peaks: low-frequency for θ = 90° and high-frequency for θ = 0°, sample №3 spectrum also has absorptivity peak at frequency of 1.4 THz, its appearance could be described by some mistakes during fabrication process. At resonant frequency of θ = 0° absorptivity of θ = 90° close to 0 and vice versa. In case of θ = 45° structure samples spectra have both low- and high-frequency peaks, but its resonant absorptivities are lower than in case of θ = 0° or θ = 90°. Table 3 presents resonant peak absorptivity for all 5 experimental samples.

Table 3. ERR geometric parameters.

Polarization angle θ, degree	Sample									
	1		2		3		4		5	
	Resonant peak frequencies, THz									
	0.396	1.053	0.326	0.866	0.277	0.746	0.242	0.651	0.219	0.581
	Resonant peak absorption, %									
0	6.1	100	4.4	99.7	4.1	83.5	9.1	89.5	7.4	85.4
45	46.6	63.4	37.5	50.4	34.4	44.9	28.7	68.9	44.1	30.6
90	98.9	6.3	50.7	1.9	68.9	3.6	78.5	1.9	60.3	2.8

6. Conclusions

Influence of incidence radiation polarization on MMA's absorptivity was investigated and following results were received:

- proposed structure is polarized sensitive with maximal peak absorptivity at polarization angles θ=0° and θ=90°
- frequencies of maximal absorptivity at θ=0° are frequencies of minimal absorptivity at θ=90° and vice versa
- polarization angles between θ=0° and θ=90° have intermediate value of absorptivity at resonant peaks

Acknowledgments
This work was financially supported by Government of Russian Federation, Grant 074-U01.

References

[1] Kuznetsov S, Paulish A, Gelfand A, Lazorskiy P and Fedorinin V 2012 *Progress In Electromagnetics Research* **122** 93
[2] Terekhov Y E, Khodzitsky M K, Grachev Y V, Sedykh E A, Belokopytov G V, Zhang X-C 2013 *Proceedings of SPIE - The International Society for Optical Engineering* ***8806*** 88062Q
[3] Gurvitz E A, Vozianova A V, Khodzitsky M K 2014 *Journal of Physics: Conference Series*, ***541***(1) 012067
[4] Denisultanov A Kh, Khodzitsky M K 2013 *Proceedings of SPIE - The International Society for Optical Engineering* **8806** 88062I
[5] Girich A, Khodzitsky M, Nedukh S, Tarapov S 2011 *NATO Science for Peace and Security Series B: Physics and Biophysics* 159-164
[6] Tarapov S I, Khodzitsky M, Chernovtsev S V, Belozorov D, Merzlikin A M, Dorofeenko A V, Vinogradov A P, Inoue M, Granovsky A B 2010 *Physics of the Solid State* ***52***(7) 1427-1431

IOP Conf. Series: Journal of Physics: Conf. Series **1062** (2018) 012013 doi:10.1088/1742-6596/1062/1/012013

[7] Tarapov S I, Khodzitskiy M, Chernovtsev S V, Belosorov D, Merzlikin A M, Vinogradov A P, Granovsky A B 2009 Solid State Phenomena ***152-153*** 394-396
[8] Duka M V, Dvoretskaya L N, Balbekin N S, Khodzitskii M K, Chivilikhin S A, Smolyanskaya O A 2014 *Quantum Electronics **44***(8) 707-712
[9] Gusev S I, Borovkova M A, Strepitov M A, Khodzitsky M K 2015 *Progress in Biomedical Optics and Imaging - Proceedings of SPIE **9537*** 95372A
[10] Borovkova M, Serebriakova M, Fedorov V, Sedykh E, Vaks V, Lichutin A, Salnikova A, Khodzitsky M 2017 *Biomedical Optics Express **8***(1) 273-280
[11] Borovkova M, Khodzitsky M, Demchenko P, Cherkasova O, Popov A, Meglinski I 2018 *Biomedical Optics Express* **9**(5) 2266-2276
[12] Tao H, Landy N, Bingham C, Zhang X, Averitt R and Padilla W 2008 *Opt. Express* **16**(10) 7181
[13] Landy N, Sajuyigbe S, Mock J, Smith D and Padilla W 2008 *Phys. Rev. Lett.* **100** 207402
[14] Landy N, Bingham C, Tyler T, Jokerst N, Smith D and Padilla W 2009 *Phys. Rev. B* **79** 125104
[15] Shen X, Cui T, Zhao J, Ma H, Jiang W and Li H 2011 *Opt. Express* **19**(10) 9401
[16] Ma Y, Chen Q, Grant J, Saha S, Khalid A and Cumming D 2011 **36**(6) 945
[17] Wang B, Zhai X, Wang G, Huang W and Wang L 2015 *Journal of Appl. Physics* **117**(1) 014504
[18] Zaitsev K, Chernomyrdin N, Kudrin K, Reshetov I and Yurchenko S 2015 *Optics and Spectroscopy* **1193** 404

IOP Conf. Series: Journal of Physics: Conf. Series **1062** (2018) 012014 doi:10.1088/1742-6596/1062/1/012014

Optical wavefront synthesis using digital holography

T A Vovk and N V Petrov

Department of Photonics and Optical Information Technology, ITMO University
49 Kronverksky Ave, Saint Petersburg 197101, Russia

E-mail: tatiana_vovk@corp.ifmo.ru

Abstract. Here we investigate the method for optical wavefront spatial formation, based on the adaptive optimization of the wavefront algorithm (AOWF). We consider the regularities of this method for the basic problems. Among the results, it is worth highlighting the possibility of creating two-dimensional optical structures that can be implemented in holographic imaging technologies.

1. Introduction

The optical beam conversion for formation of the desired spatial distribution is necessary for many applications, for example, for optical tweezers [1] and non-invasive surgical interventions using lasers [2]. Digital holography based approach to solve this problem opens the possibility to create virtual optical devices that are impossible for physical realization due to a lack of technology.

To specify the necessary amplitude distribution in a certain space region, one needs the boundary conditions that describe the wavefront in the input plane. These boundary conditions can be formed by means of holographic optical elements (HOEs) [3], as well as dynamic optical elements based on spatial light modulators (SLMs) [4], etc.

As for today, the most promising types of boundary conditions definition are the digital optical methods. These methods do not require the manufacturing of optical elements, for example, the Gershberg-Saxton iterative algorithm [5]. The development of digital optical methods for wavefront manipulation is an important stage in the development of technologies using optical radiation.

2. Statement of the problem

Numerical simulation of propagation and transformation of optical wavefronts is a known problem for optical physics. Usually, for these purposes, the paraxial approximation and scalar theory of diffraction are used.

In this paper, we investigate the method for optical wavefront formation, based on the algorithm of adaptive wavefront optimization (AOWF). Initially, this algorithm was developed to focus laser radiation through strongly scattering media [6]. We propose the application of this technique to the task of digital optical elements (DOEs). Herewith we will consider the main features of this method for the basic problems of optical physics.

3. Used methods of optical wavefront formation

In this section, we consider the implemented approaches for formation of the optical wavefronts.

3.1. Scalar diffraction theory treatment

The description of wavefront propagation can be realized through a variety of methods of the diffraction theory. We shall elaborate the method of the scalar diffraction theory, namely, the angular spectrum (AS) of plane waves. This method has several advantages over other methods. For example, it describes the optical beam propagation without introducing additional approximations. In addition, calculations through this method can be markedly accelerated by implementation of the fast Fourier transform (FFT) [7]. The AS of plane waves defines the wavefront through the Fourier transformation:

$$U(x,y) = \hat{F}^{-1}\{H(f_x, f_y, l)\cdot \hat{F}[U(x', y')]\}, \tag{1}$$

where $H(f_x, f_y, l)$ is the transfer function, given by:

$$H(f_x, f_y, l) = \begin{cases} \exp\left[i\dfrac{2\pi l}{\lambda}\sqrt{1-\lambda^2\cdot(f_x^2+f_y^2)}\right], & f_x^2+f_y^2 < \dfrac{1}{\lambda^2} \\ 0, & f_x^2+f_y^2 \geq \dfrac{1}{\lambda^2} \end{cases}, \tag{2}$$

where λ is the wavelength of the propagated optical field, and f_x, f_y are the spatial frequencies along x, y axes, respectively.

3.2. Adaptive optimization of the wavefront

Initially, the AOWF algorithm was developed to solve the problem of the beam focusing through the randomly scattering media [6]. Flat wavefront passed through a scattering sample. The resulting pattern of randomly distributed intensity maxima and minima appeared in the registration plane (see Fig. 1 (a)). After this, the AOWF algorithm was applied, which transformed the wavefront in such a way that the radiation from the source was collected at a single point on the registration plane (see Fig. 1 (b)).

An elementary case was considered earlier when there is no scattering medium in the object plane [8]. In this case, the AOWF algorithm generates an array of optimal phase delay values on the SLM, which is a discrete representation of the thin lens function (see Fig. 1 (c)). The rays of light from the source, after passing the SLM, converge at a single point on the registration plane. However, the method of target correction [6] used for the AOWF algorithm to focus into one or several focal spots is not the only possible one. If one uses more than one pixel of the matrix photodetector as a target, but a group of pixels, one can obtain a phase DOE that will focus the wavefront into a given number of points.
As a result of radiation scattering by the passage through the scattering medium, the interrelation between adjacent wavefront segments is lost in the output plane. It follows from the Huygens-Fresnel principle, which states that each point of the wavefront is a separate secondary source of spherical waves. Thus, the field in the input plane can be considered as a set of elementary radiators. Phase shift variation of radiators by means of SLM opens the possibility to configure the total wavefront in the registration plane. This fact allows to develop a simple and effective algorithm for finding the optimal configuration of elementary segments in the input plane in order to focus through the scattering medium [9].

IOP Conf. Series: Journal of Physics: Conf. Series **1062** (2018) 012014 doi:10.1088/1742-6596/1062/1/012014

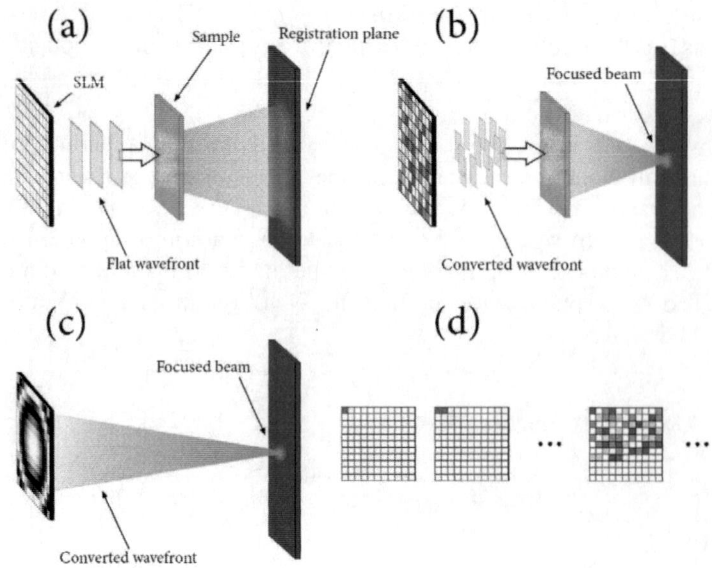

Figure 1. (a) Flat wavefront passes through a scattering sample without conversion in SLM; (b) After setting of the phase delay for each segment of the SLM, the optical beam is focused into the target image on the output plane; (c) Radiation propagation through SLM and its focusing without a scattering sample; (d) An example of the algorithm for the phase delay setting is a line-by-line search of single SLM segments.

The principle of the algorithm can be briefly described as follows (see figure 2):

1. Flat wavefront from the laser source falls onto the SLM plane;
2. Separate segments of the wavefront acquire the phase delay specified on the SLM pixels. The transformed wavefront after the passage of SLM then propagates to the registration plane;
3. The intensity distribution obtained in the registration plane is further compared by PC with the feedback condition, that is, by the distribution of the energy (target), which must be obtained in the registration plane;
4. PC selects a phase delay in the range $(-\pi, \pi]$ in the current segment (or segments) of SLM.

Then process 1 - 4 is repeated for the next segment (or segments) of SLM. There are several algorithms for selecting segments to find the optimal phase delay, for example, a line-by-line search of single SLM segments (see figure 1 (d)), binary search, or random phase masks. More details about these algorithms are given in [9].

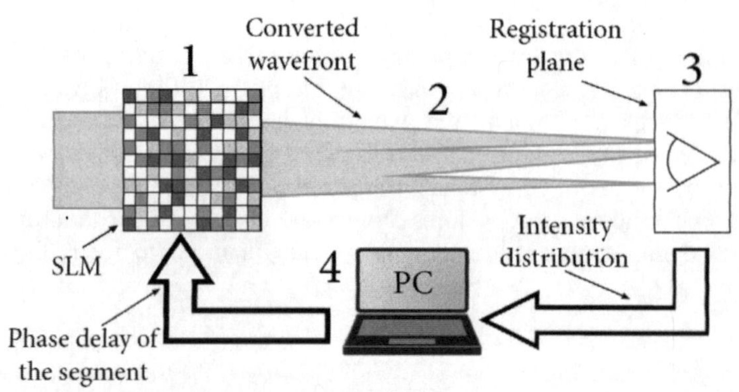

Figure 2. Algorithm of AOWF operation without a scattering medium.

IOP Conf. Series: Journal of Physics: Conf. Series **1062** (2018) 012014 doi:10.1088/1742-6596/1062/1/012014

4. Results

In this section, we analyze the basic properties of the presented method.

4.1. Formation of the point source

Let us consider the simplest case of wavefront transformation, namely, focusing in a particular point on the registration plane. The result of applying the AOWF method to obtain a point source is shown in Fig. 3.

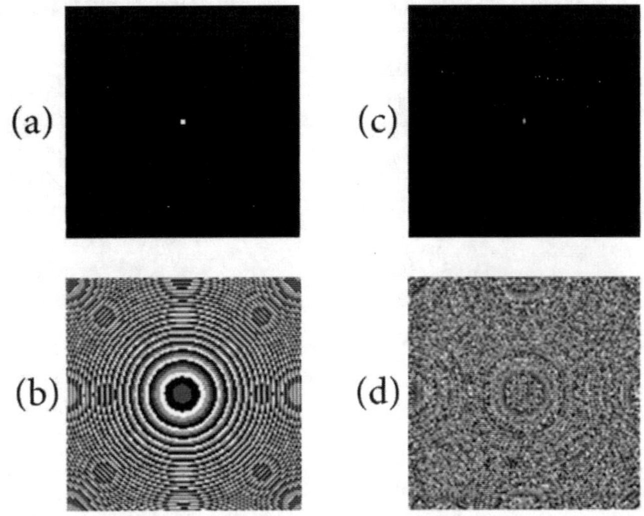

Figure 3. The target (a) and the result of the flat wavefront with homogenous intensity conversion by the AOWF method: phase (b) in the plane of SLM, amplitude (c) and phase (d) in the registration plane. Parameters: image size is 320 × 320 μm, the wavelength is λ = 532 nm, the gradation of the AOWF phase is $\varphi = 4$, the distance to the screen is $l = 2$ mm.

Uniformly distributed intensity in the plane of SLM is not shown. The wavefront from the source was redistributed to the point on the registration plane (Fig. 3 (c)). In this case, as a result of the calculation, we have the necessary phase distribution (Fig. 3 (b)) for uniform intensity (thin lens), which must be sent to SLM to obtain the required amplitude distribution.

4.2. Formation of the two dimensional image

Consider more complex intensity distribution in the plane away from the SLM. In contrast to the first problem, where the AOWF algorithm found an unambiguous distribution of phase delays for SLM, namely the function of a thin lens, with the presence of several intensity points, the solution of the problem can be ambiguous for analytical methods. The AOWF algorithm will automatically find the desired distribution, with the most uniform distribution of energy between the points of intensity. The result is shown in Fig. 4.

A beam of light with the uniform intensity distribution was redistributed into three points on the registration plane (Fig. 4 (c)). In this case, as a result of the calculation, we have the necessary phase distribution (Fig. 4 (b)), which is a sort of 'cross-linking` of phase distributions for three points separately. As seen from the comparison of Fig. 3 (c) and Fig. 4 (c), with the increase of the intensity points number undesirable additional diffraction orders appear. Methods for avoiding such effects are presented in [10].

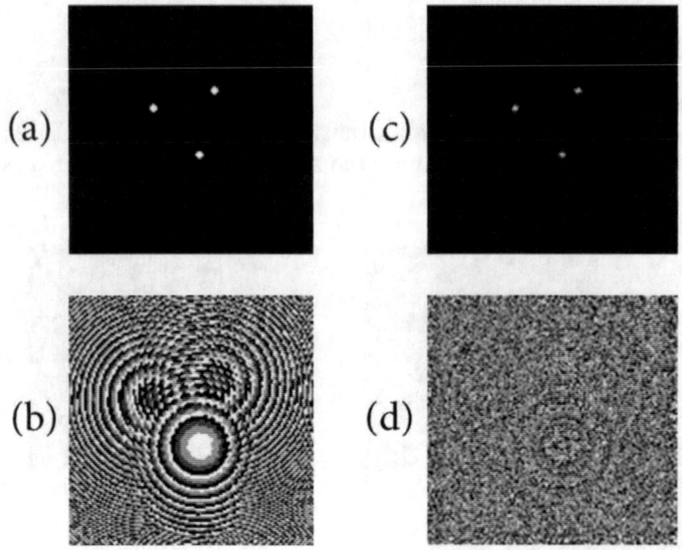

Figure 4. The target (a) and the result of the flat wavefront with homogenous intensity conversion by the AOWF method: phase (b) in the plane of SLM, amplitude (c) and phase (d) in the registration plane. Parameters: image size is 320 × 320 μm, the wavelength is λ = 532 nm, the gradation of the AOWF phase is $\varphi = 4$, the distance to the screen is $l = 2$ mm.

Let us now consider the case of the nontrivial intensity distribution in the registration plane, which is most interesting for imaging applications. In this problem, the AOWF algorithm also finds the integral phase distribution for SLM. The result of the algorithm application is shown in Fig. 5.

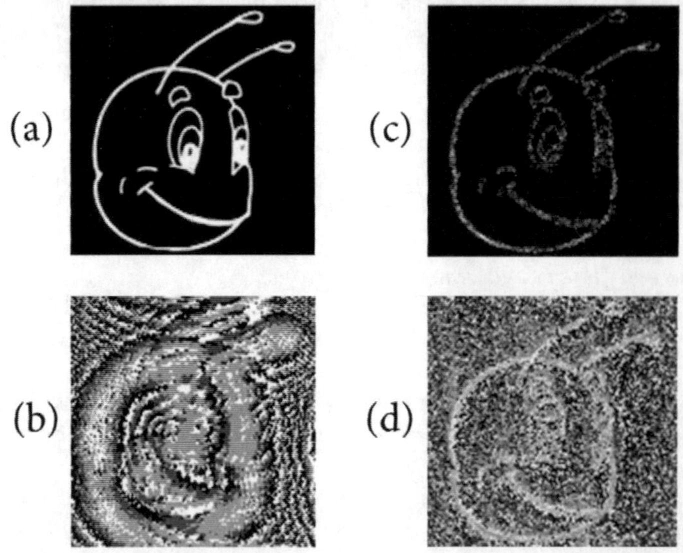

Figure 5. The target (a) and the result of the flat wavefront with homogenous intensity conversion by the AOWF method: phase (b) in the plane of SLM, amplitude (c) and phase (d) in the registration plane. Parameters: image size is 320 × 320 μm, the wavelength is λ = 532 nm, the gradation of the AOWF phase is $\varphi = 4$, the distance to the screen is l = 2 mm.

The light beam with the uniform intensity distribution was transformed into complex intensity configuration on the registration plane (Fig. 5 (c)). A large number of intensity points led to an inefficient redistribution of energy, but the threshold intensity at the points of interest is reached, and one can see the necessary object in the registration plane. Therefore, if the resulting two-dimensional projection exposes a photographic plate, it is always possible to select the necessary threshold value of the intensity and the exposure time at which the plate will be exposed only in the required region. This refinement opens the possibility of the fast and universal assignment of two-dimensional intensity distributions for holographic imaging problems.

5. Conclusion

In this paper, the method of optical wavefronts formation using adaptive wavefront optimization (AOWF) was presented and studied. The method was applied to a number of basic problems, most frequently encountered in optical physics. The obtained results showed the possibility of applying the considered method to the assignment of two-dimensional distributions of optical energy in the needed spatial area.

Acknowledgements

This work is supported by the Grant of the President of Russian Federation No. MK-7068.2016.2.

References

[1] Curtis J et al. 2002 *Opt. Commun.* **207** p 1
[2] Konig K et al. 1999 *Cell Mol Biol* **45** p 2
[3] Lukin A 2007 *J. Opt. Technol.* **74** p 1
[4] Hornbeck L 1991 Spatial light modulator (USA: pat. 4956619)
[5] Zalevsky Z et al. 1996 *Opt. Lett* **21** p 12
[6] Vellekoop I et al. 2007 *Opt. Lett.* **32** p 16
[7] Goodman J 2005 Introduction to Fourier optics (USA: Roberts and Company Publishers)
[8] Petrov N et al. 2014 Investigation of interaction of structured illumination with random scattering media (San Diego: Proc. SPIE)
[9] Vellekoop I et al. 2008 *Opt. Commun.* **281** p 11
[10] Vovk T et al. 2017 Application of digital optical phase conjugation in the problem of three-dimensional polygonal hologram formation (San Francisco: Proc. SPIE)

Polymeric recording medium Difphen – exposure features

B G Manukhin, N V Andreeva, O V Andreeva

ITMO University, 49 Kronverksky Pr., St. Petersburg, Russia

E-mail: manukhin.ifmo@gmail.com

Abstract. This paper presents experimental results of changing optical parameters of polymeric recording medium Difphen samples during exposure process. The nature of changes of the average refractive index and sample thickness is considered, quantitative estimations of spatio-temporal changes of these parameters are given.

1. Introduction

At present time, polymeric recording media are widely used for recording volume holograms which are used as spectral-angular selectors of radiation and in systems of archival storage of information. Such materials are characterized by high resolving power and information capacity of the samples in combination with available technology of their production. Their disadvantages include the fact that the change of physico-mechanical characteristics of polymer samples and their optical parameters during changing external conditions (including temperature and humidity) is significantly higher than other optical materials such as silicate glass or optical crystals. This work is devoted to research for heat effects which are accompanying the process of radiation exposure on polymeric samples during holograms recording.

2. Materials and method

Object of research: sample of material Difphen - one of the modifications of polymeric volume recording medium with diffusion enhancement based on polymethylmethacrylate (PMMA) and phenanthrenequinone (PQ) [1]. In figure 1a are shown spectral characteristics of typical sample of this material in the initial state (curve 1) and after exposure when under the action of actinic radiation there was a complete transformation of the light-sensitive molecules of PQ into molecules of photoproduct (PP), which don't have absorption in the visible region of the spectrum (curve 2).

For research of changes of optical characteristics of the sample in the process of radiation exposure and post-exposure period was used experimental stand designed for the research of phase transformations in transparent objects by method of digital holographic interferometry (DHI) [2], block-diagram of which is shown on figure 1b. On this stand an object node (3) placed outside the main scheme of the interferometer (1) which allows monitoring the process of external influence on the object by beam of radiation of 473 nm DPSS-laser (4). The sample (disk with diameter of 40 mm) is rigidly mounted in a special cassette and installed in the object beam in accordance with the optical scheme of interferometric experiment. The object beam of interferometer is a non-actinic test beam for this sample of radiation of 532 nm (figure 1a), which tests a workspace of the sample with the size not less than 25x25 mm. The beam of actinic radiation (473 nm), expanded by lens, passes through the sample in the center of workspace.

IOP Conf. Series: Journal of Physics: Conf. Series **1062** (2018) 012015 doi:10.1088/1742-6596/1062/1/012015

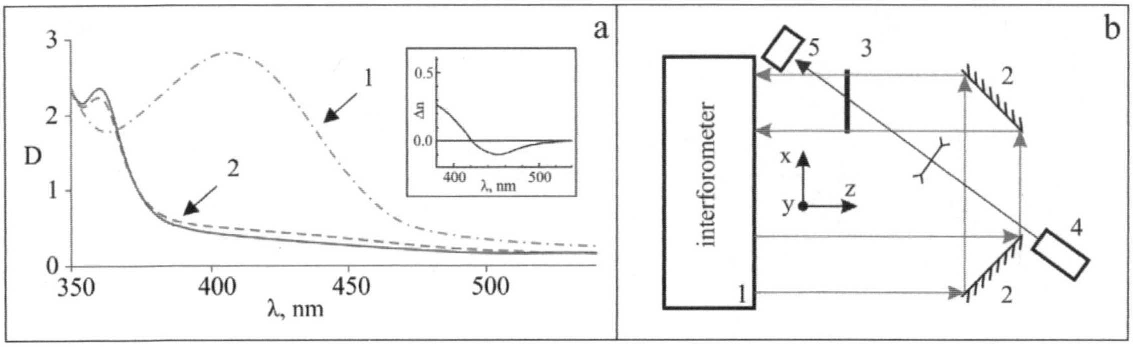

Figure 1. a – spectral characteristics of sample of material Difphen before radiation exposure (curve 1) and after complete cycle of processing (curves 2); b – block-diagram of experimental stand: 1 – interferometer, 2 – mirrors, 3 – object, 4 – laser (473 nm), 5 – detector.

During radiation exposure that is used for holograms recording, happens the absorption of energy. Part of this energy goes to heating of the sample and part to photochemical conversion of PQ into PP with a quantum yield of ~50%. Heating leads to reversible changes of optical thickness, the sign and order of magnitude of which is pre-theoretically impossible to estimate. The amount of change of the optical constants of the samples during photochemical conversion is not experimentally researched in the visible region of the spectrum, however, as theoretical calculations with use of the Kramers–Kronig relations for medium PMMA/PQ have shown [3], this dependence in the visible region of the spectrum has a complex character (figure 1a, insert). Observed during radiation exposure phase changes of the sample are the sum of reversible and irreversible changes and depend on the level of exposure, time of exposure, wavelength and power of exposed radiation. In the post-exposure period observed phase changes are caused only by photochemical processes of conversion PQ into PP.
Thus, changes of the optical thickness during radiation exposure are caused by photochemical and heating processes and in the post-exposure period only by photochemical processes. Therefore, considered stage of exposure and post-exposure stage, we could estimate separately the contributions of heating and photochemical processes in the observed phase changes of the sample. This type of experiment is used in this work.

3. Results and discussion

Experimental results for making exposure dependence of changes of optical characteristics of the material Difphen and for identifying of the magnitude of contributions of reversible and irreversible components of these changes are obtained on typical sample of material Difphen with a thickness of 2.6 mm with the same power density of the exposed radiation, with a wavelength of 473 nm, but with different time of exposure. Four series of similar measurements with exposure of 10 s (series 1), 30 s (series 2), 200 s (series 3), 600 s (series 4) with radiation with a power of 23 mW with a beam diameter of 8 mm were carried out. The sequence of digital holograms was recorded with an interval of one second and with exposure time 1/1024 s. Registration began 10 seconds before the exposure and ended after 300 seconds after it. During exposure in each experimental series the control of the power of exposed radiation transmitted through the sample was carried out.
On figure 2 obtained results are given in the form of exposure dependencies of changes of optical thickness of the sample (curves 1 and 2), and comparison of irreversible, $\Delta(nl)^{PP/PQ}$, and reversible, $\Delta(nl)^{T}$, changes that accompany the process of exposure of the sample (curve 3) is given.

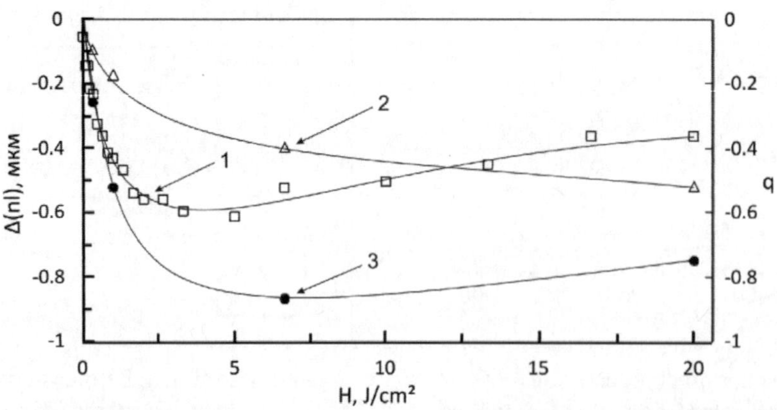

Figure 2. Exposure dependencies of changes of optical parameters of the sample Difphen with a thickness of 2.6 mm at power density of exposed radiation of 33 mW/cm^2 , with wavelength of 473 nm: 1 - $\Delta(nl)^{\Sigma}$; 2 - $\Delta(nl)^{T}$; 3 – q = $\Delta(nl)^{PP/PQ}/\Delta(nl)^{T}$.

Typical area of optimal exposures for holograms recording on the material Difphen is not exceed the value of 2 J/cm^2, exactly in this area of exposures in the current experiments reversible changes in several times higher than the value of irreversible phase changes of the sample.

Table. The change of the refractive index of polymeric medium, due to the conversion of PQ into PP (at 532 nm).

№	Exposure, *H*, J/cm^2	Δn, 10^{-5}
1	0.33	1.0 ± 0.4
2	1.0	3.3 ± 0.6
3	6.6	13.1 ± 0.7
4	19.8	15.23 ± 0.14

In assumption that the geometric thickness of the sample after exposure has not changed relative to the thickness of the sample in the initial state (or these changes are negligible compared to the change of refractive index), were determined changes of the refractive index of the sample, due to its discoloration (conversion PQ into PP) during different levels of exposure, are shown in table. The obtained results in order of magnitude are in agreement with data of work [4].

4. Conclusion

In this work experimental data on the change of optical characteristics of samples of polymeric recording material Difphen under the influence of exposed radiation with a wavelength of 473 nm is presented.

Quantitative estimation of phase changes of the samples carried out with a temporal resolution of 1 s by method of digital holographic interferometry during exposure and post-exposure period, allowed us to separate the contributions of heating and photochemical processes in change of optical thickness of the samples.

The exposure dependence of change of the refractive index of the material Difphen, due to the conversion of PQ into PP is received and it is shown that during conversion of PQ into PP under the influence of radiation of 473 nm the value of these changes during monitoring with wavelength of reading $\lambda = 532$ nm is the value of $\Delta n \sim 1.5 \cdot 10^{-4}$.

References

[1] Andreeva O V, Bandyuk O V, Paramonov A A, Cherkasov A S, Gavrilyuk E R and Andreev P V 2000 *J. Opt. Tech.* 67(12) p 1043

[2] Manukhin B G, Gusev M E, Kucher D A, Chivilikhin S A and Andreeva O V 2015 *Opt. Spectros.* **119(3)** pp 392–397

[3] Veniaminov A V 2012 Dynamics of photoinduced gratings in polymeric materials: influence of molecular diffusion *Doctoral Dissertation* (S.I. Vavilov State Optical Institute)

[4] Lumeau J and Lequime M 2006 *Appl. Opt.* **45(24)** pp 6099–6105

Peculiarities of rainbow holograms application on the surface of the caramel

A P Toropova, I Y Santalina and M I Fokina

ITMO University, 49 Kronverksky Pr., St. Petersburg, 197101, Russia

E-mail: mfokina@niuitmo.ru

Abstract. Rainbow hologram may be applied on the candies made from isomalt, fructose, sugar and molasses. The composition of caramel affects on its optical properties and the ability to save the diffraction grating.

1. Introduction

Everyone knows from childhood what caramel is. This amazing candy has various shapes, it is transparent and colored like glass. We assumed that the new aesthetic appearance of a traditional confectionery product can be decorated with a rainbow hologram. The rainbow hologram allows to create bright images with a color that depends on the viewing angle, due to the diffraction relief. Researches on the application of rainbow holograms on the surface of confectionery products were carried out earlier, but received wide practical application not yet [1-4]. The goal of this research was the creation of a diffraction grating on caramel. Caramel can be made from different materials - sugar, molasses, fructose, isomalt. Isomalt is a substitute for sugar, which has become popular to confectioners last time.

Sugar is a food product of sweet taste, made from sugar cane or sugar beet. Sugar exists in the form of colorless monoclinic crystal blocks, has a melting point t= 185 ° C, is soluble in water and easily crystallizes in supersaturated solutions. Sugar is stable in the air, but hygroscopic, it is able to absorb the moisture of the air [4,5].

Molasses is a sugary product made by hydrolysis of starch with enzymes or dilute acids followed by filtration and boiling syrup. Molasses is a mixture of glucose, oligosaccharides and dextrins in water. The content of solids in molasses is about 80%. Molasses is transparent and colorless in pure form. It increases the solubility of sucrose and blocks crystallization, that is why it is used in caramel production [6].

Fructose is a monosaccharide, found in many fruits in a free form, and as a part of oligosaccharides and polysaccharides. Fructose is extracted from fruits by hydrolysis of sucrose and inulin, as well as the action of alkalis on glucose. Fructose has a melting point t = 102 ° C, is highly soluble in water, hygroscopic to a large extent [7].

Isomalt is a sugar substitute in the form of white crystals, flavourless with a sweet taste, its sweetness is 42% of the sweetness of sucrose. It is a mixture of hydrogenated monosaccharides and disaccharides. Isomalt has a melting point t = 150 ° C, it is readily soluble in water, hygroscopic to a small extent [8,9].

IOP Conf. Series: Journal of Physics: Conf. Series **1062** (2018) 012016 doi:10.1088/1742-6596/1062/1/012016

2. The experimental results

Three types of caramel samples were prepared for the research. The first sample was made by melting the isomalt crystals about temperature of 80 ° C, followed by a sharp heating till 150 ° C. The solid transparent colorless sample was obtained after cooling. The second sample was obtained by melting fructose crystals about temperature of 80 ° C, followed by a sharp heating till 100°C. The curved transparent sample with a yellowish shade was obtained after cooling. The third sample was prepared as follows: sugar was dissolved in water about temperature of 90 ° C, molasses was added, the mixture was cooked till 120°C. The solid transparent sample with a yellowish shade was obtained after cooling.

The transmission coefficients of light were measured on three samples with a thickness of 4 mm for estimating the optical properties of the caramel. The graphs of the dependence of the transmittance on the wavelengths are presented in figure 1.

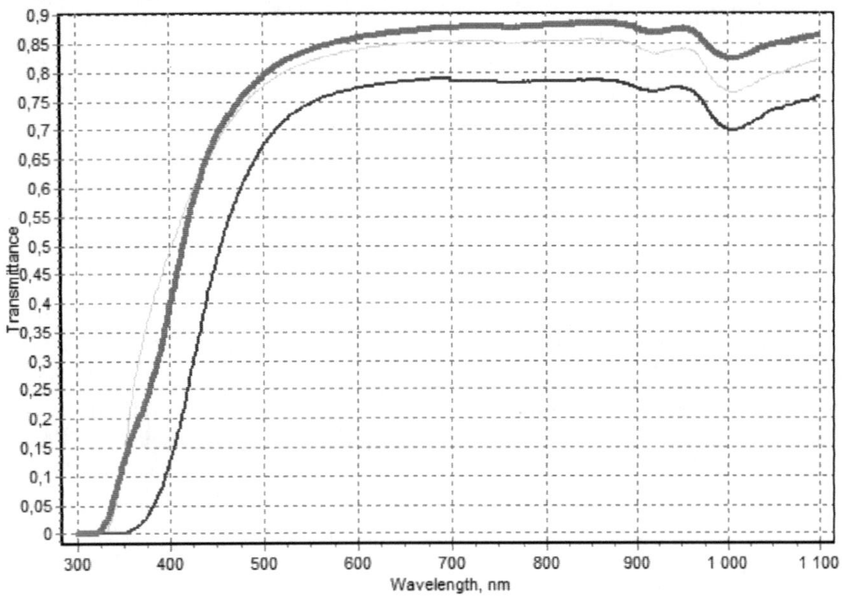

Figure 1. Graph of the dependence of the transmittance on the wavelength for samples from isomalt (green narrow line), fructose (blue middle line), mixtures of sugar, molasses and water (red bold line).

According to the graphs caramel samples are non-transparent for the ultraviolet part of the spectrum and are transparent in the visible part of the spectrum, the transmittance of caramel from fructose is about 80%, from isomalt is about 85%, from the mixture of sugar, molasses and water is about 90%.

Two diffraction gratings were selected for the research - with a period of 20 lines per mm and 600 lines per mm. Silicone molds were made from this gratings, caramel was poured into molds. The samples were taken out of the mold after solidification and examined for the presence of a diffraction grating. Figure 2 shows the photo of diffraction of laser radiation on a caramel sample with a diffraction grating with a period of 20 lines per mm. Figure 3 shows the photo of the diffraction of white light on a caramel sample with a diffraction grating with a period of 600 lines per mm. The bright rainbow effect can be observed on a grating with a period of 600 lines per mm.

IOP Conf. Series: Journal of Physics: Conf. Series **1062** (2018) 012016 doi:10.1088/1742-6596/1062/1/012016

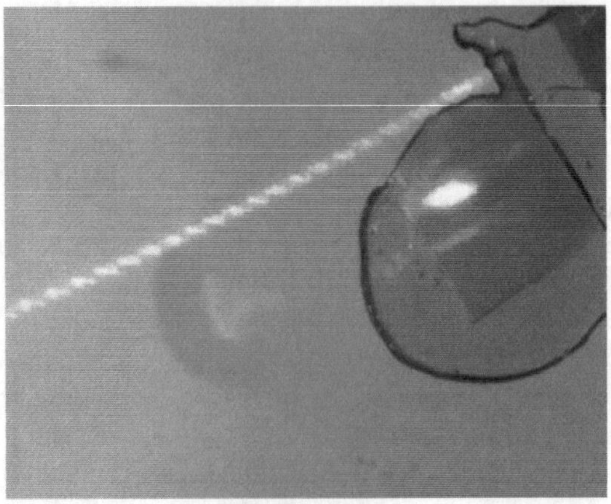

Figure 2. Diffraction of laser radiation on a caramel sample with a diffraction grating with a period of 20 lines per mm.

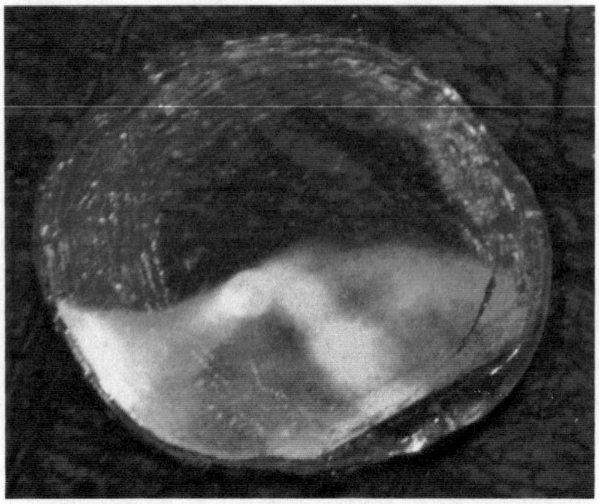

Figure 3. Diffraction of white light on a caramel sample with a diffraction grating with a period of 600 lines per mm.

It was found, that the lifetime of the diffraction grating is different for the investigated caramel samples. The lifetime of diffraction grating on sample of fructose was about 5 minutes. The rainbow effect became less bright after 12 hours, and after 24 hours the diffraction grating disappeared on the sample of isomalt. A sample of a mixture of sugar, molasses and water initially had a bright rainbow across the surface of the sample, after 17 hours the rainbow became less bright, after 24 hours, diffraction was observed on a narrow part of the sample, after 27 hours the diffraction grating disappeared.

The diffraction grating was saved longer on a sample of sugar, molasses and water than on other samples. Molasses can be different types and it is differ by hygroscopicity. Further research was continued with different molasses, the following three samples were made. The first sample was prepared from sugar, water and starch maltose molasses. The second sample is cooked from sugar, water and corn syrup. The third sample was made from sugar, water and starch caramel molasses. The lifetime of the first sample was 27 hours, the second - 20 hours, the third - 34 hours. Thus, the diffraction grating was stored longest on the sample with a starch caramel molasses.

Caramel is viscous-flowing substance at room temperature, so when it is removed from the mold, it tends to take a shape with the lowest surface energy, as a result of this the diffraction grating disappeared. In addition, relaxation processes are possible in the volume of caramel after its removing from the mold, associated with the release of residual water. One of ways of solution this problem is addition of structuring agent to the amorphous structure of caramel. As a result, it can be turned into a three-dimensional structure more resistant to external influences, temperature and humidity. Agar-agar was chosen as such substance. Agar-agar is a product obtained from red algae, having the form of a powder or flakes from white to yellow colour, used as a gelling agent, stabilizer, thickener, water-retaining agent, clarifier, film former, agent for encapsulation. Agar-agar is readily soluble in boiling water and insoluble in cold water. Jelly from agar-agar melts at a temperature not lower than 80 ° C [8, 10]. The sample was made from sugar, water, starch caramel molasses and agar-agar. The solid transparent yellow shade sample was obtained after cooling, the diffraction grating on its sample disappeared after 40 hours.

The estimation of diffraction efficiency was presented for caramel samples from sugar, molasses and water in Table 1. The most diffraction efficiency is obtained sample of sugar, water and corn syrup, but the lifetime of the diffraction grating of its sample was the least. Despite the small values of diffraction efficiency, the rainbow effect is clearly discernable visually (see figures 2, 3).

IOP Conf. Series: Journal of Physics: Conf. Series **1062** (2018) 012016 doi:10.1088/1742-6596/1062/1/012016

Table 1. Diffraction efficiency of caramel samples

№	Composition	Diffraction efficiency,%
1	Sugar, water, starch maltose molasses	0,05
2	Sugar, water, corn syrup	0,13
3	Sugar, water, starch caramel molasses	0,005
4	Sugar, water, starch caramel molasses, agar-agar	0,06

3. Conclusion

It was concluded from the experiments that caramel is capable of transmitting the relief, but it is not capable of save it for a long time. The least stable material is fructose, and the most stable is caramel from sugar, starch caramel molasses, water and agar-agar. The life time of the samples reaches 40 hours, it is comparable to the expiration date of confectionery products containing creamy, custard, curd, yoghurt creams. Thus, caramel decorations with a rainbow hologram can be used as indicators of expiration date and observance of storage conditions, for example, cakes.

References

[1] Begleiter E 1987 *Holographic Products* p 4
[2] Begleiter E 1991 *Proceedings of the SPIE* **1461** pp 102–109
[3] Begleiter E 2004 *Edible articles that include edible optical elements and methods for producing same* p 21.
[4] Ponce-Lee E L, Olivares-Pérez A and Fuentes-Tapia I 2004 *Optical Materials* **26** p 6
[5] Rabinovich V A and Khayavin Z Ya 1978 *Brief Chemical Handbook. Edition 2, revised and supplemented.* (Leningrad branch "Chemistry") p 316
[6] Molasses. Great Soviet Encyclopedia. Electronic resource [access mode: http://bse.sci-lib.com/article087410.html; date of circulation: 20/09/2016; Russian language].
[7] Fructose. Great Soviet Encyclopedia. Electronic resource [access mode: http://bse.sci-lib.com/article117739.html; date of circulation: 20/09/2016; Russian language].
[8] Sarafanova L A 2004 *Nutritional Supplements: Encyclopedia. The second edition, corrected and supplemented.* (St. Petersburg: GIORD) pp 276–278
[9] Korneeva O S and Bozhko O 2005 *Fundamental research* **10** pp 34–35
[10] Lazerson I I and Sokiryansky F L 2012 *Culinary science or scientific culinary* (Moscow: Tsentrpoligraf) p 150

Chirped volume Bragg grating recording in photo-thermo-refractive glass

A E Angervaks, S A Ivanov, S I Klimentev and R A Okun

ITMO University, 49 Kronverksky pr., St. Petersburg, 197101 Russia

angervax@mail.ru

Abstract. Volume chirped Bragg gratings (volume CBG) recording optical scheme is implemented and calculated. Chirped gratings are recorded in photo-thermo-refractive (PTR) glass and can be used in near-infrared (near-IR) spectral range. The set of elements of optical scheme is optimized in order to realize recording of chirped Bragg gratings with various characteristics using limited set of elements.

1. Introduction

Currently, holographic optical elements (HOE) play an increasing role in a various fields of science and technology connected with managing and/or conversion of optical radiation. In the first place it is due to progress in creation and development of holographic media. Thus, volume holographic media are of great demand for production of spectral and spatial selectors, combiners/decombiners of optical beams, narrow-band optical filters, Bragg mirrors for increasing of spectral radiance and thermal stabilization of emission wavelength of semiconductor lasers [1, 2].

Physical properties of bulk crystalline and glass holographic materials enable recording of chirped Bragg grating (CBG) within them. Chirped grating is a volume 3-D hologram with its period variation along the direction of the beam propagation (along the grating vector). In this grating Bragg condition for different wavelengths of incident radiation is occurred at appropriate depth inside the grating.

The main functions of chirped grating are stretching of ultra-short laser pulse in time to amplify it (spectral chirp acquisition) and subsequent pulse compression to duration close to initial one. Stretched pulses have lower power density and can be amplified without damage of elements of amplifying system. This technique called chirped pulse amplification (CPA) was proposed in [3]. Initially, pulse stretching and compression in CPA systems were carried out almost exclusively by pairs of surface diffraction gratings [4, 5]. But such stretchers and compressors are bulky, difficult to align, and sensitive to vibrations.

An important advancement in the development of CPA systems was made by the use of fiber-chirped Bragg gratings [6, 7]. Their use enabled creation of much more simple and flexible schemes. But limited aperture of chirped fiber Bragg gratings lead to limitations on the peak power achievable with fiber-based pulse compressors.

To increase the peak power of stretched/compressed laser pulses the use of volume chirped Bragg gratings (volume CBG) was proposed [8]. The sufficient progress in formation and implementation of volume CBGs is strongly connected with the development of a new bulk holographic material – photo-thermo-refractive (PTR) glass, which enables recording of high-efficient volume holograms [9-

IOP Conf. Series: Journal of Physics: Conf. Series **1062** (2018) 012017 doi:10.1088/1742-6596/1062/1/012017

11]. The use of PTR-based stretchers and compressors decreased the size and weight of amplifying optical systems, enhanced the robustness of CPA systems to effects of vibration and shocks [12–15]. Volume CBGs recorded in photo-thermo-refractive glass can be used in the spectral range from 0.4 to approximately 2.5 μm with diffraction efficiencies exceeding 90%. They provide pulse stretching up to 1 ns and compression down to 200 fs for laser pulses with energies and average powers exceeding 1 mJ and 250 W, respectively, while keeping the recompressed beam quality M2 < 1.4 [16].
In the present paper calculation of volume CBG recording optical scheme is carried out. Chirped gratings are recorded in PTR glass produced in ITMO University and intended for use in near-IR spectral range. Set of cylindrical lenses of optical scheme is optimized in order to realize recording of chirped Bragg gratings with various characteristics. Using the designed calculation algorithm, one can calculate feasible combinations of lenses and spatial parameters of the optical scheme for volume CBGs recording with preassigned characteristics.

2. Properties of PTR glass

The fluoride PTR glass was designed and synthesized in ITMO University, Russia [17]. The fluoride PTR glass is a photosensitive multi-component sodium–zinc–alumino-silicate one containing fluorine (6 mol.%) and a small amount of bromine (0.5 mol.%) also doped with additives (cerium, antimony and silver) that are responsible for the photo-thermo-induced precipitation of silver nanoparticles and sodium fluoride crystals. Untreated fluoride PTR glasses are transparent in a wide spectral range of 250–2500 nm. The selective UV irradiation (325 nm) into the Ce^{3+} absorption band in the spectra of these glasses results in the formation of neutral silver molecular clusters. The subsequent heat treatment of UV-irradiated PTR glasses near the glass transition temperature (T_g) induces the silver nanoparticle formation. The thermal treatment of these glasses at temperatures above T_g leads to the growth of silver bromide shell on a silver nanoparticle [18] and then to the precipitation of sodium fluoride cone on it [17, 19].
Photo-thermo-induced crystallization in a form of NaF microcrystals growth within the glass volume leads to the local refractive index modulation (Δn). Nowadays, the maximum refractive index modulation amplitude (RIMA) for the fluoride PTR glass can be as high as 1.5×10^{-3}. If scattering by the crystalline phase inside the glass is not critical, the maximum RIMA magnitude can be even greater (like 3.5×10^{-3}). Volume Bragg gratings recorded in these glasses reveal a unique combination of working characteristics such as the high angle and spectral selectivity, high diffraction efficiency, high mechanical and optical strength, and high thermal and chemical durability [17].

3. Optical scheme for volume CBG recording

In volume CBG the condition of Bragg diffraction appearance for different wavelengths at variation of the refractive index $n_{eff}(z)$ or/and the period of the grating $d(z)$ along the direction of the beam propagation is as follows:

$$\lambda_{Bragg} = 2n_{eff}(z)d(z) \tag{1}$$

In the present paper the Bragg condition (1) is realized via variation of the grating period, $d(z)$, with the refractive index, $n_{eff}(z)$, remaining constant.
Recording of volume CBG in the PTR glass is produced in a standard transmission two coherent-wave mixing scheme (figure 1). The system of cylindrical lenses is included in each recording channel to form the shape of the recording beam's wavefronts – the first one is convergent and the second one is divergent. Points S_0 и R_0 are located symmetrically relatively to y axis and at the same distances from z axis. Such scheme enables formation of parallel planes of CBG along its length within PTR glass volume.
Experimental optical scheme for CBG recording based on the concept, illustrated in figure 1, is shown in figure 2. Recording laser beam is formed by collimator *8* and illuminates the beamsplitter *1*.

Cylindrical lens *5* forms diverging beam with desired focal length. In the second arm of the recording scheme lenses *4* and *6* form converging beam with the same focal length.

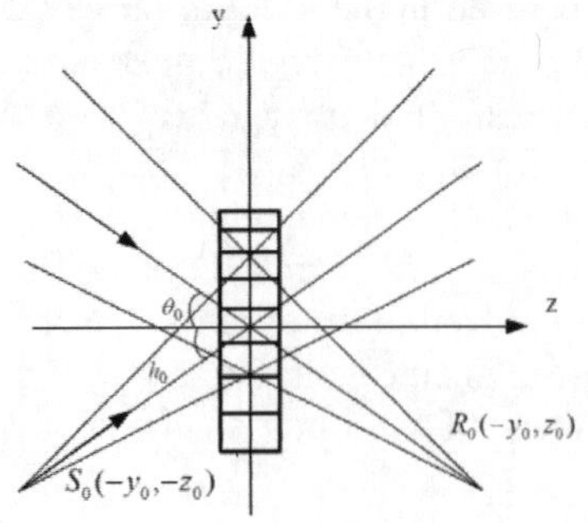

Figure 1. Volume CBG recording method [20].

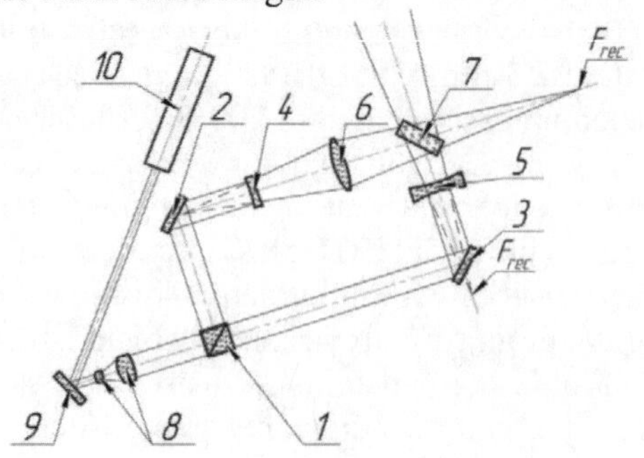

Figure 2. Experimental scheme for volume CBG recording in PTR glass. *1* – beamsplitter, *2, 3* – plane mirrors, *4, 5* – negative cylindrical lenses, *6* – positive cylindrical lens, *7* – sample of PTR glass, *8* – laser beam collimator, *9* – plane mirror, *10* – recording laser. F_{rec} – points similar to S_0 and R_0 ones in figure 1.

4. Optical scheme parameters calculation

In order to calculate parameters of the optical scheme in figure 2 (focal lengths of lenses, linear distances between elements), the following initial data were used:

1. center wavelength of CBG (λ, nm);
2. spectral width of CBG ($\Delta\lambda$, nm);
3. CBG length (l, mm);
4. He-Cd recording laser wavelength – 325 nm (λ_{rec});
5. diameter of the recording beam at the beamsplitter *1* – 1–20 mm (d_{laser});
6. refractive index of PTR glass – 1.498 (n).

On the basis of λ and $\Delta\lambda$ values maximum (d_{max}), middle (d_{mid}) and minimum (d_{min}) periods of CBG were calculated.

$$d_{max,\,min} = \frac{\lambda \pm \frac{\Delta\lambda}{2}}{2n} \tag{2}$$

$$d_{mid} = \frac{\lambda}{2n} \tag{3}$$

Corresponding convergence angles of the recording beams are as follows

$$\theta_{max,mid,min} = \arcsin\left(\frac{\lambda_{rec}}{2d_{min,mid,max}}\right) \tag{4}$$

According to convergence angles and CBG length values focal lengths of the recording beams, f_{rec}, were calculated (these are the distances from the center of the sample *7* to F_{rec} points – see figure 2)

$$f_{rec} = \frac{l\cos(\theta_{mid})}{\sin(\theta_{max} - \theta_{min})} \tag{5}$$

Using formulae (2)–(5) focal length values of lenses *4–6* (f_4, f_5, f_6) and optical diameter of lens *6* (d_6) were calculated (diameters of lenses *4* and *5* are determined by the diameter of laser beam and should be less or equal to the value of 20 mm). Initial diameter of the recording beam (d_{laser} = 10 mm), distance from the imaginary focus of lenses *4* and *6* to the lens *6* ($L_{4\text{-}6}$ = 1300 mm) and distance between elements *6* and *7* ($L_{6\text{-}7}$ = 300 mm) were taken under the assumption of calculation convenience but may be varied slightly.

$$d_6 = l\cos\left(\theta_{mid}\right)\frac{f_{rec}+L_{6-7}}{f_{rec}} \tag{6}$$

$$f_4 = L_{4-6}\frac{d_{laser}}{d_6} \tag{7}$$

$$f_5 = f_{rec}\frac{d_{laser}}{l\cos\left(\theta_{mid}\right)} \tag{8}$$

$$f_6 = \left(\frac{1}{L_{4-6}}+\frac{1}{f_{rec}+L_{6-7}}\right)^{-1} \tag{9}$$

Calculations using formulae (5)–(9) were carried out for the central wavelength of CBG equal to 1053, 1064, 1550 nm, spectral width of 10–30 nm and CBG length of 10–30 mm. Analysis of the resulting array of parameters showed that the maximum number of CBGs with desired characteristics can be recorded using the following optimal set of focal lengths of lenses *4–6* – see table 1.

Table 1. Optimal set of lenses for CBG recording.

Element	Focal length (mm)	
Lens *4*	500	1000
Lens *5*	500	2000
Lens *6*	500	1000

Lenses *4* and *5* are negative cylindrical with diameter not less than 30 mm; lens *6* is a positive cylindrical with the diameter not less than 38 mm. One can calculate the ranges of the main CBG parameters that can be realized using different combinations of lenses from table 1 (see table 2).

Table 2. Feasible lenses combinations and corresponding CBG parameters.

f_4 (mm)	f_5 (mm)	f_6 (mm)	λ (μm)	l (mm)	$\Delta\lambda$ (nm)
500	500	500	1.0–1.5	2–30	3.6–120
500	500	1000	1.0–1.5	2–50	3.6–80
1000	500	1000	1.0–1.5	2–27	3.6–81
500	2000	1000	1.0–1.5	2–25	0.9–24
1000	2000	1000	1.0–1.5	2–30	0.9–40

Using above deduced formulae one can estimate all variants of lenses combinations and linear dimensions of the optical scheme for the recording of CBG with preassigned characteristics. It can be useful in situation of the limited number of optical elements (especially cylindrical lenses). As an example, we made such calculation for the CBG with the following parameters: λ = 1053 nm, l = 30

IOP Conf. Series: Journal of Physics: Conf. Series **1062** (2018) 012017 doi:10.1088/1742-6596/1062/1/012017

mm, $\Delta\lambda$ = 30 nm, f_{rec} = 1790 mm. In theory, several combinations of lenses from table 2 can be used for the recording. But calculation showed that the unique set of lenses (f_4 = 500 mm, f_5 = 500 mm, f_6 = 1000 mm) provides the diameter of lens *6* and linear distances between elements of the optical scheme that can be realized in practice. Using of all other sets leads to negative or extremely large values of linear distances and cannot be implemented.

5. Conclusions

Optical scheme for volume CBGs recording in PTR glass is proposed and calculated. Calculations were carried out for CBG central wavelength from near-IR spectral range. The set of optical elements is optimized to cover the wide range of CBG's parameters. Formulae deduced allow estimating all variants of lenses combinations and linear dimensions of the optical scheme for the recording of CBG with preassigned characteristics. It can be useful in situation of the limited number of optical elements.

Acknowledgments

This work was financially supported by Ministry of Education and Science of Russian Federation in the framework of state task #16.1651.2017/PCh.

References

[1] Volodin B L, Dolgy S V, Melnik E D, Downs E, Shaw J and Ban V S 2004 *Opt. Lett.* **29** 1891
[2] Glebov L 2013 *Rev. Laser Eng.* **41** 684
[3] Strickland D and Mourou G 1985 *Opt. Commun.* **56** p 219
[4] Treacy E 1969 *IEEE J Quantum Electron.* **5** p 454
[5] Martinez O 1987 *IEEE J Quantum. Electron.* **23** p 59
[6] Boskovic A, Guy M J, Chernikov S V, Taylor J R and Kashyap R 1995 *Electron. Lett.* **31** p 877
[7] Galvanauskas A, Fermann M E, Harter D, Sudgen K and Bennion I 1995 *Appl. Phys. Lett.* **66** p 1053
[8] Galvanauskas A and Fermann M E 1996 *U.S. Patent* 5,499,p 134
[9] Stookey S D, Beall G H and Pierson J E 1978 *J. Appl. Phys.* **49** p 5114
[10] Panysheva E I, Tunimanova I V and Tsekhomskii V A 1990 *Fiz. Khim. Stekla* **16** p 239
[11] Kuchinskii S A, Nikonorov N V, Panysheva E I, Savvin V V and Tunimanova I V 1991 *Opt. Spektrosk.* **70** p 1286
[12] Liao K H, Cheng M Y, Flecher E, Smirnov V I, Glebov L B, Galvanauskas A 2007 *Opt. Express* **15** p 4876
[13] Glebov L B, Flecher E, Smirnov V I, Galvanauskas A and Liao K-H 2008 *U.S. Patent* 7,424,185
[14] Frankinas S, Michailovas A, Rusteika N, Smirnov V, Vasilyeu R and Glebov A L 2016 *Proc. SPIE* **9730** 973017
[15] Nejbauer M, Kardaś T M, Stepanenko Y and Radzewicz C 2016 *Optics Letters* **41** p 2394
[16] Glebov L, Smirnov V, Rotari E, Cohanoschi I, Glebova L, Smolski O, Lumeau J, Lantigua C and Glebov A 2014 *Opt. Eng.* **53** 051514
[17] Ivanov S A, Ignatiev A I, Nikonorov N V and Aseev V A 2014 *Journal of Optical Technology* **81** p 356
[18] Glebov L B, Nikonorov N V, Panysheva E I, Petrovskii G T, Savvin V V, Tunimanova I V and Tsekhomskii V A 1992 *Opt. Spectrosc.* **73** p 237
[19] Dyamant I, Abyzov A S, Fokin V M, Zanotto E D, Lumeau J and Glebova L N 2013 *J Non–Cryst. Solids* **378** p 115
[20] Do D–D, Kim N, Kwon O–W, An J–W, Jeon S–H and Lee K–Y 2007 *Proc. SPIE* **6488** 648814

Fabrication and properties of elements based on volume transmission gratings with a surface relief

P P Sokolov and N D Vorzobova

Saint Petersburg National Research University of Information Technologies, Mechanics and Optics, St. Petersburg, Russia

E-mail: vorzobova@mail.ifmo.ru

Abstract: The fabrication processes, diffraction and selective properties of volume transmitting holographic gratings with a surface relief are considered. The conditions for obtaining structures with a diffraction efficiency of the volume component up to 80% and a relief component up to 45% at angles of incidence up to 50°, as well as possible practical applications of elements based on such structures are determined.

The processes of fabrication and properties of elements based on volume transmission gratings are of great practical interest and are being widely studied at present due with the possibilities of obtaining high diffraction efficiency and high angular selectivity that determine their using in information storage devices and for obtaining light beams formation elements and selective elements. One of the perspective directions is the fabrication of elements by the holographic method in photopolymer materials. A distinctive feature of this work is the orientation on obtaining elements on the basis of volume gratings with a surface relief. Such gratings have not been previously investigated and have properties that determine new possibilities for using elements based on them. In previous works [1,2] was observed the formation of the surface relief on a volume grating with refractive index modulation, but the height of the relief was insignificant and did not significantly affect the diffraction properties of the structures.

In this paper the problems of determining the conditions for obtaining a volume grating with refractive index modulation with the greatest diffraction efficiency, the formation of a relief grating on its surface with a significant depth of the relief, and investigating the diffraction properties of such structures were solved.

The experiments were performed with holographic recording of transmission gratings at a wavelength of 442 nm at a power density of $3 \cdot 10^{-2}$ W/cm^2. As recording materials, acrylate monomer compositions based on commercial components [3] are used. The layers were formed and exposed between two surfaces - a substrate and a coating film. Surface relief was formed by removing the coating film and post-exposure processing in isopropanol. Diffraction efficiency was determined at a wavelength of 650 nm as the ratio of the intensity in the first order of diffraction to the intensity of the incident radiation.

IOP Conf. Series: Journal of Physics: Conf. Series **1062** (2018) 012018 doi:10.1088/1742-6596/1062/1/012018

To determine the conditions for obtaining an effective volume grating, the dependence of the diffraction efficiency on the thickness of the layer and the recording frequency was investigated. It has been found that the greatest diffraction efficiency of 80% is achieved for thicknesses of about 100 μm and recording frequencies of about 300 l/mm (figure 1) with an exposure time of 40 sec. The gratings are volume, which is confirmed by the availability of only one order of diffraction, and also by the Klein criterion [4]. For a grating with thickness of 100 μm and a period of 3 μm, Q = 29.

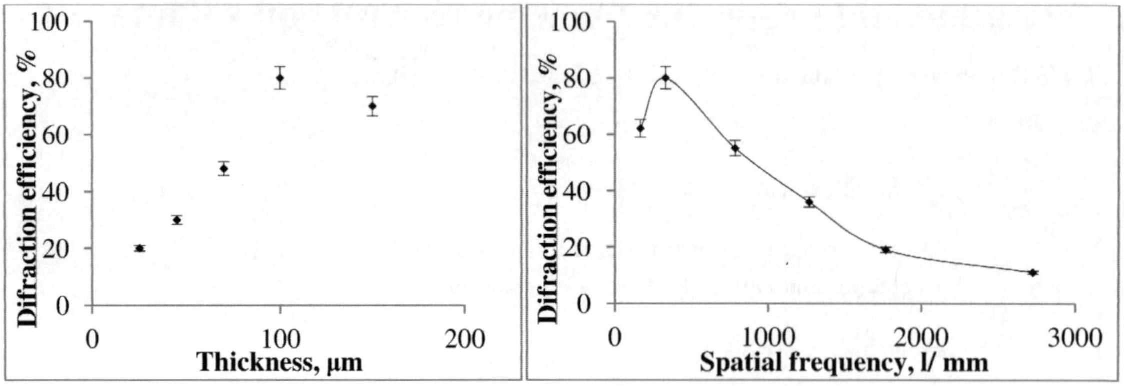

a b

Figure 1. Dependence of the diffraction efficiency on the layer thickness (a) and the recording frequency (b).

It is established that when the coating is removed after exposure, on a surface grating a relief grating is formed. To determine the conditions for obtaining a grating with the greatest depth of the relief, the dependence of the diffraction efficiency of the relief grating on the duration of exposure was studied. It is established that a lattice with a significant depth of relief is formed in a very narrow range of exposure parameters (Table 1).

Table 1. Dependence of the maximum diffraction efficiency of the relief grating on the exposure duration.

№	Exposure duration, s	Difraction efficiency, %
1	20s	20
2	40s	45
3	1 min	35
4	2 min	22

A relief grating with a maximum diffraction efficiency of 45% can be considered as quasi-volume, because this value exceeds the DE limit for a thin grating (33%). In addition, calculation of the diffraction efficiency using Kogelnik's theory [5] for a relief height of 0.3 μm obtained for this grating by atomic force microscopy gives a close value - of 46%.

The study of the angular selectivity of volume gratings with a surface relief - hybrid structures showed a significant broadening of the angular selectivity contour relative to the volume grating. Figure 2 shows the contour of the angular selectivity of the hybrid structure obtained at the optimal conditions in comparison with the angular selectivity of the volume grating. It can be seen that high diffraction efficiency is maintained when the deviation from the Bragg angle for the volume component of the hybrid structure is up to 50°.

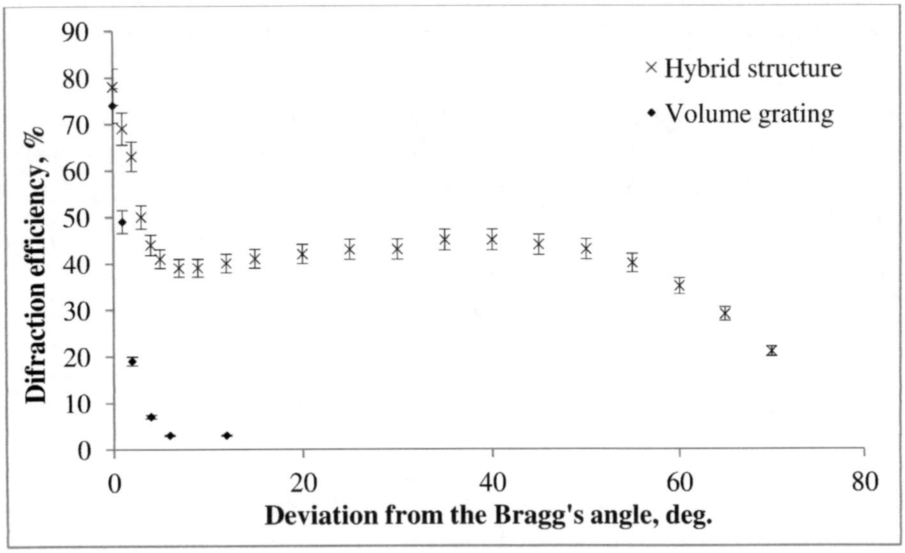

Figure 2. Change in diffraction efficiency for hybrid structure by deviation from the Bragg's angle.

In the figure 3 it can be seen that the diffraction properties of a volume grating with a surface relief are conserved at a large angle of incidence of radiation from a white light source.

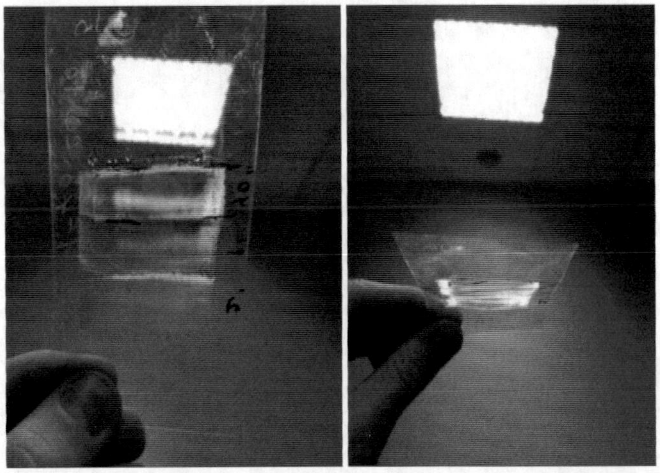

Figure 3. Diffraction on hybrid structure with a change in the angle of incidence.

The established diffraction and selective properties of volume gratings with a surface relief determine possible applications of elements based on them. One of the possible applications is the fabrication of protective elements. Elements based on hybrid structures make it possible to observe the image over a wide range of incidence angles when the maximum brightness is achieved in a narrow range of angles, which is an advantage in technologies based on visual control over elements based on volume gratings. Elements based on hybrid structures can also be used as graphic design elements. Another possible application is the use in solar energy technologies to produce elements of solar concentrators that allow the use of solar radiation in a wide range of incidence angles. It is possible to use one-dimensional hybrid structures operating in one plane and as well as two-dimensional structures that can be obtained with multibeams interference. Elements based on volume gratings with a surface relief have advantages relative to elements based on either volume or relief gratings [6-8] associated with increasing the range of incidence angles, reducing the number of elements and simplifying the fabrication process.

References

1. Naydenova I, Mihaylova E, Martin S, Toal V 2005 *Optics Express* **13**, 4878-89
2. Lucchetta D E, Spegni P, Di Donato A, Simoni F, Castagna R, 2015 *Opt. Mater.* **42,** 366–9.
3. Yu.E. Burunkova Yu E, S. A. Semina A, L. N. Kaporski L N, and V. V. Levichev V V 2008 *J. Opt. Technol.*, **75**, 653-7.
4. Colier R J, Burckhardt C B, Lin L, 1973 *Optical Holography*, Academic Press, NY, London, 686 p.
5. Kogelnik H 1969 *Bell Syst. Tech. J.* **48,** pp 2909–47
6. Jong T, Boer D, Bastiaansen C 2011 *Opt. Expess* **19,**15127 -15142
7. Akbari H, Naydenova I, Martin S, 2014 *Appl. Opt*. **53**, 1343-63
8 J.H. Lee, H.Y. Wu, M.L.Piao, and N.Kim 2016 *IEEE Photon. J.* **8**, 1-11.

Estimation of optical heterogeneity of samples in the process of developing nanoporous matrices from two-phase glass

N V Andreeva, A O Ismagilov and O V Andreeva

ITMO University, 49 Kronverksky Pr., St. Petersburg, Russia

E-mail: andreevanv_3@mail.ru

Abstract. In this paper the research of optical heterogeneity of nanoporous silicate matrices (NPM) by method of digital holographic interferometry (DHI) is presented.

1. Introduction

Currently, the increase of complexity of devices and expanding of the area of application of optical range radiation sets new requirements to optical materials. One of the main ways of development is the creation of new materials with specified properties. Matrix principle of construction of composite materials is one of the prospective.

Nanoporous silicate matrices have a special place among porous structures [1]. At present time, application of nanoporous silica matrices mainly related to the transparency of these glasses in the visible spectral range and to the possibility of obtaining samples of optical quality [2].

The research of nanoporous matrices occupies a certain place in the life of science community. To obtain a material which will possess the specified properties it is important to develop technology that will include comprehensive researches for estimation of quality of the product.

The aim of this work was to research the influence of each stage of chemical processing on the final optical heterogeneity of the sample.

2. Materials and method

In the experiment we used nanoporous silicate matrices that were made of sodiumborosilicate two-phase glass formed by two interpenetrating phases: chemically unstable borate and chemically stable silica. We used matrices in the form of polished disks with a diameter of 15 mm and with a thickness of 1 mm and in the form of plane-parallel plates, which were made using proven technology from two-phase glass DV-1 after stage of heat processing.

The main characteristics of the samples during their development are the average pore diameter and the free pore volume, characterizing the volume of the sample that isn't occupied by silica frame. The average pore diameter of the samples is 17 nm. The free pore volume of such samples is in the range of 48-58% and depends on the time of the procedure of alkaline etching. During developing nanoporous silica matrices technological regulation was observed and it gave an ability to reproduce the specified characteristics of NPM from batch to batch.

Control of phase homogeneity of the samples was carried out at each stage of processing by carrying out researches of samples on the stand of DHI – scheme of the stand (figure 1) - in the initial state, after acid etching and after alkaline etching. Digital interferograms of the samples were obtained and then they were analyzed.

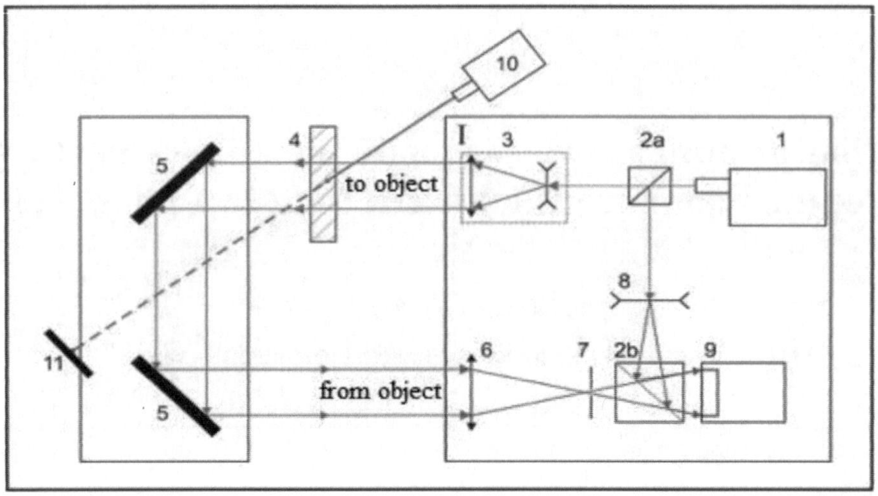

Figure 1. Experimental stand for the research of transparent objects by method of digital holographic interferometry: 1 - the probe laser; 2a, 2b - beamsplitter cubes; 3 - collimator; 4 - object of research; 5 - rotary mirror system; 6 - lens 1; 7 - aperture diaphragm; 8 - lens 2; 9 - CMOS-matrix; 10 - DPSS laser (532 nm); 11 - device OPHIR.

An additional problem of the research was to determine the influence of gravity on the result of the procedure of chemical processing. During setting to a chemical processing the samples were oriented so that the action of gravity relative to the sample was always pointing in the same direction. In order to determine the influence of gravity on the result of the etching before alkaline etching sample 1 was rotated on 180° relative to the horizontal axis and sample 2 was left in the same position.
As the main parameter for estimation of optical heterogeneity was used the effective refractive index n_{eff} and its changes throughout the volume of the sample.
The effective refractive index is determined as following:

$$n_{eff} = V_1 n_1 + V_2 n_2, \quad (1)$$

where V_1 – the relative free pore volume; n_1 – the refractive index of the immersion of free pore volume (air immersion – n_1=1, water immersion – n_1=1.33); V_2 – the relative volume of sample that is occupied by the frame SiO_2; n_2 – the refractive index of the sample frame (n_2=1.45).
Calculation of the free pore volume was carried out by the weight method which was proposed in the following work [3].
The effective refractive index of the sample calculated by the formula 1 represents by itself a refractive index averaged throughout the volume of the sample.
In table 1 the parameters of the obtained matrices during filling of pores with air and water after each stage of chemical etching are shown.

Table 1. Parameters of the obtained matrices during filling of pores with air and water after acid (stage 1) and after alkaline (stage 2) etching.

Sample	Stage 1			Stage 2		
	V_1,%	$n_{eff}(n_1 = 1)$	$n_{eff}(n_1 = 1.33)$	V_1,%	$n_{eff}(n_1 = 1)$	$n_{eff}(n_1 = 1.33)$
1	22.56	1.35	1.42	54.31	1.20	1.38
2	22.64	1.34	1.42	53.88	1.20	1.38

IOP Conf. Series: Journal of Physics: Conf. Series **1062** (2018) 012019 doi:10.1088/1742-6596/1062/1/012019

3. Results and discussion

Interferograms characterizing the phase structure of the initial samples were obtained. The samples were set in a cuvette which was included in the stand of DHI.

Table 2. Phase structure of billets for disks samples which was measured in the air (left) and in the water (right).

Sample	In the air	In the water
1		
2		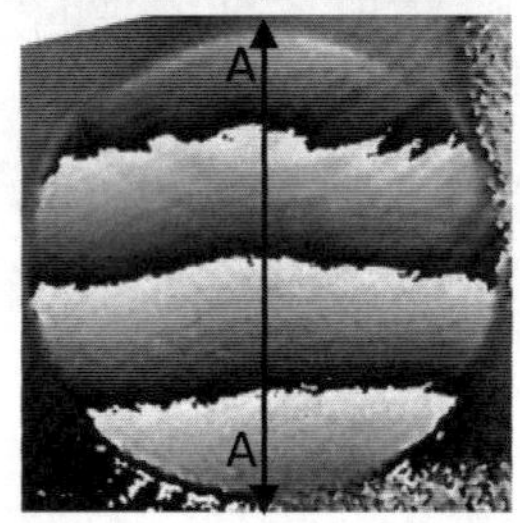

Optical heterogeneity of the samples after chemical processing due to the uneven of distribution of the free pore volume throughout the volume of the sample. Meanwhile the thickness of the sample 1 doesn't change. Estimation of uneven of the initial billets of samples 1 and 2 (column 3 of table 2) and samples 1 and 2 after all stages of chemical etching (column 6 of table 3) in water immersion (n_l=1.33).

Interferograms characterizing the phase portrait of the sample are shown in table 2 for the initial state of researched disks sample before chemical etching. In this case, the observed interference fringes with the same period *d*, indicate about insignificant wedging of the samples (dotted line on figure 2).

Interferograms of samples 1 and 2 after each stage of etching are shown in table 3 and they show optical heterogeneity – the distance d between two adjacent fringes is changed in the vertical direction (section A-A coinciding with the direction of gravity), this suggests that these changes are due to the influence of gravity on diffusion chemical processes.

The distance between two interference fringes corresponds to a phase difference of 2π. Thus, in the centre of the sample changes of phase of 2π happens on the section $\Delta z = d = 5$ mm, and at the top of the sample on the section $d = 0.2$ mm, where d is the distance between two interference fringes. It is the value $\Delta n/d$, which characterizes the optical heterogeneity of the sample, the effective refractive index which, as a rule, is averaged throughout the volume.

IOP Conf. Series: Journal of Physics: Conf. Series **1062** (2018) 012019 doi:10.1088/1742-6596/1062/1/012019

Table 3. Phase structure of samples after two stages of technological processing.

Sample	After stage 1 in the air	After stage 1 in the water	g (z)	After stage 1 in the air	After stage 1 in the water	*Δn/d*, mm^{-1}
	Top	Top		Bottom	Bottom	
1	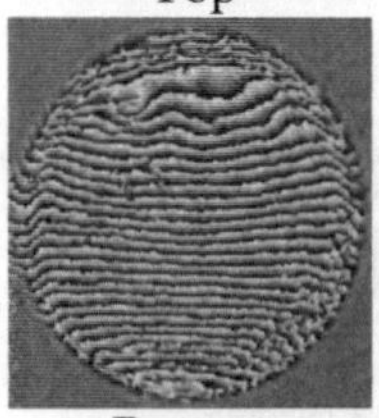	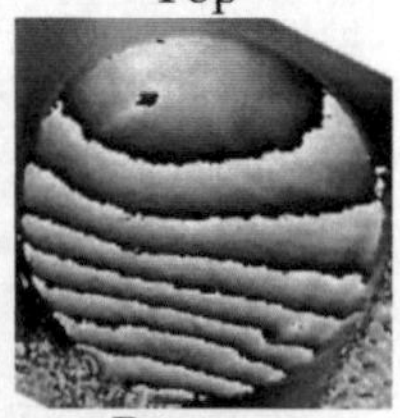	↓	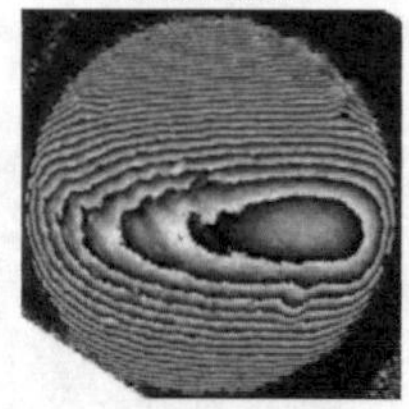	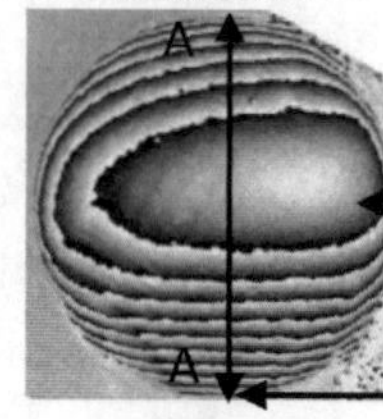	8*10^{-5} 18*10^{-4}
	Bottom	Bottom		Top	Top	
	Top	Top		Top	Top	
2	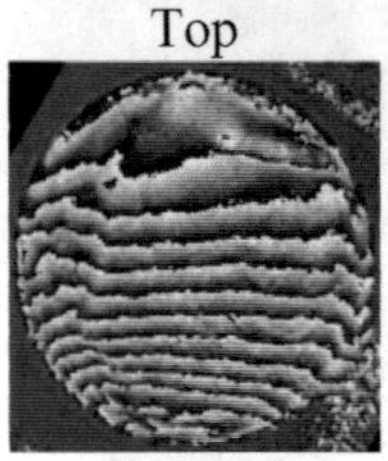		↓	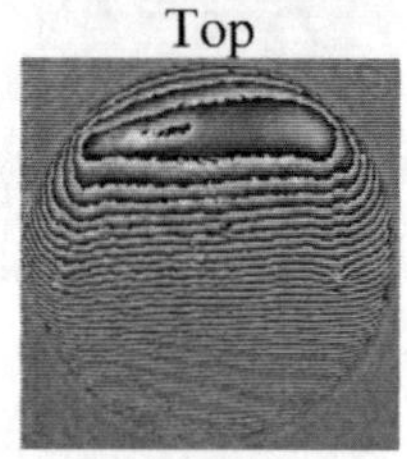	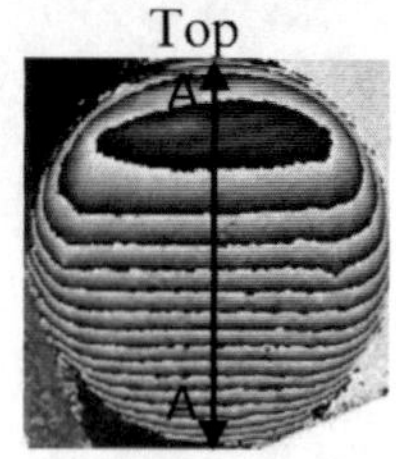	
	Bottom	Bottom		Bottom	Bottom	

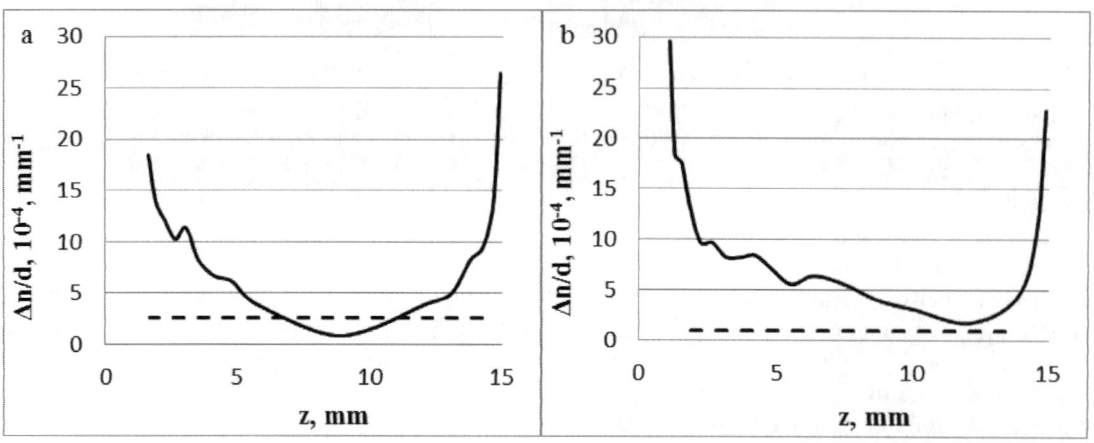

Figure 2. The change of the value *Δn/d*, in the section of samples 1 and 2 along the line A-A (table 2 and 3), dotted line – billets in water, solid lines - the samples after all stages of chemical etching in water.

4. Conclusion

Optical heterogeneity of the samples due to the influence of chemical etching is associated with the removal of the soluble phase that is enriched with the oxides of boron and sodium during acid etching and with the removal of "secondary" silica from the area of destruction of borate phase during alkaline etching.

The change of the optical heterogeneity is associated with the orientation of the samples in the gravity field and its value is 8x10^{-5} in the center and 18x10^{-4} on the edge of the sample 1. During turning of the sample before alkaline etching there is a shift to the center, while in the sample that wasn't turned over there is only a gain of heterogeneity obtained during acid etching.

IOP Conf. Series: Journal of Physics: Conf. Series **1062** (2018) 012019 doi:10.1088/1742-6596/1062/1/012019

References

[1] Andreeva O V, Bandyuk O V 2011 *Holograms – Recording Materials and Applications* ed Naydenova I (Rijeka: InTech) pp 45–70
[2] Khoo I–C et al. 1994 *Series in Nonlinear Optics* vol 2 *Nonlinear Optics and Optical Physics* (Singapore: World Scientific) p 481
[3] Girsova M, Drozdova I and Antropova T 2014 *Glass Physics & Chemistry* **40(2)** pp 162–166

IOP Conf. Series: Journal of Physics: Conf. Series **1062** (2018) 012020 doi:10.1088/1742-6596/1062/1/012020

Study of diffusion in bulk polymer films below glass transition: evidences of dynamical heterogeneities

V V Lesnichii[1,2], **A V Veniaminov**[2] **and E Bartsch**[1]
[1] Institute of Physical Chemistry, Albertstraße 21, Institute of Macromolecular Chemistry, Stefan-Meier-Str. 31, University of Freiburg, Freiburg im Breisgau 79104, Germany
[2] Department of Photonics and Optical information technologies, ITMO University, Kronverksky pr, 49, Saint-Petersburg 197101, Russia

E-mail: vasilii.lesnichii@physchem.uni-freiburg.de

Abstract. By means of the Forced Rayleigh Scattering (holographic grating relaxation) technique the diffusion of dye molecules in bulk films of polymer below entanglement molecular weight was investigated around the glass transition temperature. In the temperature range [T_g–20K; T_g + 5K] a splitting of the dye diffusion coefficient into two separate branches was revealed. The phenomenon was found to be similar to that reported for thin polymer films, which may prove the dominating role of dynamic heterogeneities in the investigated temperature region.

1. Introduction

Various techniques are being used in polymer physics to probe material properties in the vicinity of the glass transition temperature (T_g), as well as for studies of the glass transition process itself: differential scanning calorimetry (DSC) and gel permeation chromatography (GPC), dynamic mechanical analysis, dielectric spectroscopy and others. However, these conventional methods are not always sensitive enough to local effects that are related to e.g. segmental motion of macromolecules. One of possible explanations of the nature of the glass transition relies on coexistence of spatial domains in polymer melt that differ in extent of order [1]. The Forced Rayleigh Scattering (FRS) technique using photoattachable dye markers emerges to be a powerful instrument for monitoring local and collective movements of polymer species with respect to spatial microinhomogeneity. In the present work poly (para-methylstyrene) (PpMS) doped with phenanthrenequinone (PQ) has been employed as model system for such investigations.

2. Materials and methods

PpMS was chosen for its similarity in physical characteristics to the extensively studied polystyrene. The additional methyl group attached at the para- position allows photochemical attachment of PQ molecules to polymer chains [2–4], thus providing an opportunity to study both macromolecular and tracer diffusion in a conventional FRS experiment [5,6].

2.1. Sample preparation and characterization

The polymer was kindly prepared by the group of Dr. Lutz at the Institute Charles Sadron, Strasbourg, using anionic polymerization. The molecular mass distribution (M_w = 15 kg/mol ≈ M_c/2, PDI = 1.06)

was controlled by size exclusion chromatography and GPC. 13C-NMR study proved the atactic nature of the polymer. DSC (10 °/min) measurement was employed to acquire the T_g value of 106 °C.
In order to achieve a homogeneous distribution of the photoreactive dye phenanthrenequinone (PQ) in the polymer (1.5% to polymer mass); PQ was added to a PpMS solution ($C \approx C^*$) in benzene and then freeze dried for 6 hours in liquid nitrogen in the dark. The resultant flakes were ground manually to thin powder in a mortar. The powder was then placed in the metal cuvette between two quartz windows and moulded with gradual compression above the glass transition but below the decomposition temperature (160 °C) for at least 12 hours, resulting in clear quartz-polymer sandwich of about 100 μm thickness. Keeping in mind the crucial effect of time-temperature history on glassy polymers, temperature was carefully controlled in the course of consequential measurements; no steep changes of temperature were applied, and after each switching to a new temperature (e.g. from 100 to 95 °C), the cuvette with the sample was allowed to equilibrate for at least one week before the measurements started.

2.2. Forced Rayleigh Scattering

The FRS, or holographic (grating) relaxation, technique [7–9] may be considered from either a holographic or scattering point of view. The ranges of diffusion coefficients D and spatial scales Λ on which the latter can be measured are $D = 10^{-10} - 10^{8}$ $\mu m^2 s^{-1}$ and $\Lambda = 10^{-1} - 10^{2}$ μm, respectively, which makes FRS ideal for slow diffusion measurements in polymer systems even below T_g. The smaller spatial period Λ the more sensitive is the technique to slow motion. Therefore reflection gratings [9] were used to study extremely viscous glassy polymer. Among the experimental techniques that need concentration gradients to detect diffusive mass transfer, a special group consists of those utilizing periodic spatial distributions of concentration (gratings) that can be easily created and monitored by optical means. The idea of diffusion studies with periodic structures can be demonstrated with one-dimensional Fickian equation of diffusion solved under harmonic initial conditions. The solution shows that the cosinusoidal shape of concentration profile is conserved, while its modulation amplitude obeys simple exponential decay. A grating itself can be produced in a light-sensitive material by imprinting an interference pattern of two beams, and its modulation can be monitored by measuring the diffraction efficiency. This allows for retrieval of diffusion coefficients.

2.3. Phenanthrenequinone and complementary grating effect

Depending on the wavelength of the laser and the absorption spectrum of the dye in its intact and its photoexcited (phototransformed) states, either phase, amplitude, or (in general) a mixed amplitude-phase grating can be formed in response to the interference pattern. Photosensitive molecular tracers responsible for holographic recording can react to the radiation in different ways, including photophysical excitation, heating, or photochemical transformation. While photochromic dyes undergo a trans-cis photoisomerization or, as e.g. fulgides, ring opening that leads to the shift of the absorption band and a corresponding change in refraction index sufficient for efficient hologram recording [10], PQ molecules form semiquinone radicals, which attack any adjacent methyl group (e.g. on polymer chains) and form covalent bonds with the environment (macromolecules, if they surround the photosensitive species). Photoattachment itself leads to slight changes in absorption and refraction; therefore, only weak diffraction is observed immediately on exposure. However, important is that the recording of the interference pattern gives birth to two gratings, shifted with respect to each other by half a period: one formed of the phototransformed dye molecules (ACR with opened ring, or attached PQ) and another – of intact molecules (closed ring or free PQ). Both those patterns imprinted into the medium will relax (dissipate) because of diffusion of the corresponding species. Therefore the relaxation of the observable diffraction efficiency may appear nonmonotonous. The phenomenon known as complementary grating effect [11] is especially pronounced if the diffusivity of the dye changes dramatically on phototransformation. The extreme example is provided by PQ and its photoproduct whose diffusion coefficients differ by several orders of magnitude, thus making grating growth and decay caused by diffusion of respectively PQ molecules and PQ-labelled macromolecules

IOP Conf. Series: Journal of Physics: Conf. Series **1062** (2018) 012020 doi:10.1088/1742-6596/1062/1/012020

well separated in time [5]. This is why polymeric systems doped with PQ are convenient for studying probe and chain diffusion in one single experiment as well as for recording high-efficiency stable volume holographic elements [6,12].

2.4. FRS setup

Although the readout can be done with a laser of any wavelength if the corresponding Bragg conditions are satisfied, the simplest setup arrangement presumes this to be done by one of the two beams that were used for exposition. In such a case Bragg conditions are satisfied automatically, since the angle of incidence and the wavelength both remain unchanged, while the intensity can be reduced by introducing a filter in order to prevent further transitions of dye molecules. The description of the setup is presented in a caption to the figure 1.

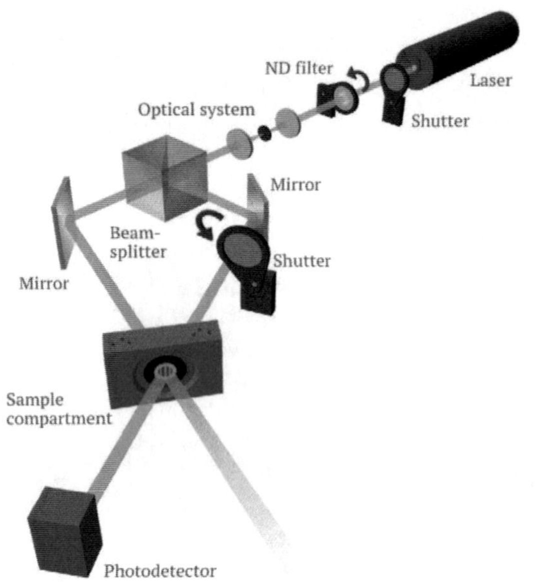

Figure 1. Scheme of FRS experimental facility used in this work. The coherent linearly polarized beam of a CW 0.6 W DPSS laser ("Torus" from Laser Quantum, $\lambda = 532$ nm, attenuated to 40 mW operating power) is spatially filtered by optical system and then split 50:50. On the recording stage both beams are overlapping at the sample compartment, where the polymer film is fastened vertically in a thermally regulated cuvette ($\Delta T = 0.1$ °C). Varying the angle between the two interfering beams changes the grating distance from the minimum value accessible in counter-propagating scheme to essentially unlimited large values at almost co-propagating beams, thus enabling one not only to tune relaxation time to convenient duration of experiments but also properly choose the microscopic spatial scale of experiment, a feature especially important in the case of anomalous diffusion.

During the readout stage one of the split beams gets blocked by the shutter and a ND filter attenuates the reading beam such that no further phototransformation of dye occurs. Continuous diffracted light is then collected by the photodetector in logarithmically arranged time slices of 100 ms duration.

3. Results

In figure 2, typical relaxation kinetics of gratings recorded in PpMS and measured at temperatures below glass transition are presented in the form of best fits with the stretched biexponential equation

$$I(t) = \left\{ A\exp\left[-\left(\frac{t}{\tau_1}\right)^{\beta_1} \right] + B\exp\left[-\left(\frac{t}{\tau_2}\right)^{\beta_2} \right] \right\}^2,$$

which degenerates into a single exponential above T_g. As one can immediately see, these graphs relate to the stage of the gratings' reveal (see section 2.2.1) and thus are considered as representing the diffusion of a free PQ dye. Moreover, there are two processes, both the characteristic times and contributions of which depend on temperature: "fast" and "slow". In the temperature range investigated the slow process is suppressed by the fast one than more, than higher the temperature is.

The diffusion coefficients corresponding to the two processes can be plotted versus temperature to track the Williams-Landel-Ferry (WLF) and Arrhenius dependencies, as shown in figure 3. The diffusion of small molecules in glass-formers as function of temperature below the glass transition is usually represented by activational Arrhenius mechanism, while diffusion of probes above T_g is described by WLF equation. The relatively big discrepancy of coefficients for "slow" process around 85°C in the area of low temperatures may be explained as a result of insufficient thermal stabilization (one week) and is in the focus of the ongoing work. However, recent studies of thin polymer films have shown similar behaviour of relaxation times below the T_g ([13], figure 3).

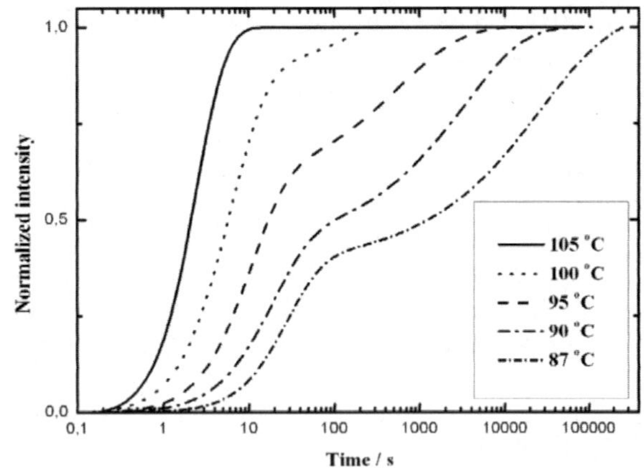

Figure 2. Normalized diffracted light intensity vs. time. Reflection gratings with 180 nm spacing were consequently recorded in different spots of the PQ-doped PpMS sample and read out at five different temperatures (consequentially downwards).

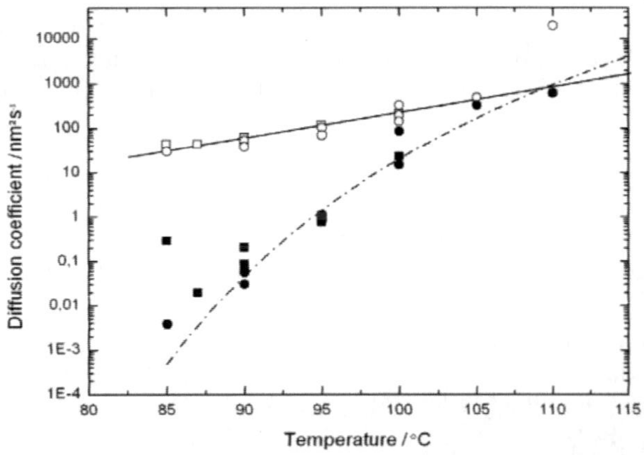

Figure 3. Diffusion coefficients of the fast (open symbols) and slow (full symbols) processes as functions of temperature. The solid line corresponds to the Arrhenius fit with activation energy of 144 kJ/mol. The dash-dotted line is the WLF-fit. Squares correspond to the series of measurements made consequently downwards upon temperature range, circles – vice versa.

4. Discussion of results

The interaction between polymer and solvent depends on temperature since the polymer-solvent interaction parameter (χ-value) does so; a good solvent may become bad and vice versa, if temperature changes. This means, for example, that polymer chains unentangled in solution will shrink to coils while cooling down in the beginning of freeze drying procedure intended for making a solid powder out of solution. Moreover, the following evacuation inevitably leads to the substitution of however good solvent with vacuum which is the worst "solvent" possible. This makes polymer coils shrink due to van-der-Waals attraction. In case the solution was not entangled in the beginning of freeze drying (that is, at concentrations below the critical overlap concentration, C*), this will lead to a so-called "reentanglement dynamics" [14] in the polymer films prepared out of the freeze dried polymer flakes, which can take much more time than a standard theory of reptation predicts. In general, there are plenty of articles aiming at the question whether a polymer can retain its configuration from solution being freeze dried [15–19]. It is worth mentioning that some similarities were found by FTIR studies in the behaviour of freeze-dried spin-cast thin polymer films [20]. The explanation of these similarities is driven by the idea that fast solvent removal peculiar to both preparation techniques leaves polymer chains in a similar non-equilibrium state.

5. Conclusion

In this study the dynamics of dye molecules in the melt of short polymer chains ($M < M_c$), which was obtained by freeze drying its unentangled solution, was investigated below the glass transition temperature. It was revealed that the bifurcational behaviour of the diffusion coefficient takes place in this temperature region, which indicates the coexistence of domains with two distinct mobilities. While the nature of one type of domains, with smaller diffusion coefficients, is inherited from the polymer structure above glass transition temperature, the second demonstrates activational mechanism of the dye's movement, as if dye molecules would be trapped within specifically ordered structures formed by polymer chains. At the same time, the diffusion coefficients for this branch of D(T) behaviour are higher, than for the WLF-branch. This gives rise to the explanation that below T_g there are dynamical ordered structures with relatively short time of existence – dynamical heterogeneities. Although these heterogeneities are commonly supposed to be of the order of few nanometres closely above T_g [21], some other studies for glass-forming liquids have shown the presence of domains up to hundreds of nanometres [22]. This, as well as the reasonable expectation of heterogeneities to grow upon moving below glass transition explains our ability to sense them with FRS. The reason these heterogeneities induce in bulk samples similar dye dynamics as in thin polymer films may lie in sample preparation technique. Further investigations with model systems prepared from solutions with higher concentrations and higher molecular weights are in the focus of the ongoing work.

References

[1] Asai M, Shibayama M and Koike Y 2011 *Macromolecules* **44** 6615–24
[2] Carapellucci P A, Wolf H P and Weiss K 1969 *J. Am. Chem. Soc.* **91** 4635–9
[3] Maruyama K, Otsuki T and Naruta Y 1976 *Bull. Chem. Soc. Jpn.* **49** 791–5
[4] Kim S S et al. 1996 *Bull. Korean Chem. Soc.* **17** 983–4
[5] Veniaminov A V and Sillescu H 1999 *Macromolecules* **32** 1828–37
[6] Veniaminov A V, Bartsch E and Popov A P 2005 *Opt. Spectrosc.* **99** 744–50
[7] Eichler H J, Günter P and Pohl D W 1986 *Laser-induced dynamic gratings* (Berlin, Heidelberg, New York, Tokyo: Springer)
[8] Antonietti M, Ehlich D, Sillescu H and Wesselman M 1989 *Prog. Colloid Polym. Sci.* **80** 83–92
[9] Veniaminov A V and Bartsch E 2005 *Opt. Spectrosc.* **101** 290–8
[10] Kardinahl T and Franke H 1995 *Appl. Phys. A* **61** 23–7
[11] Park S, Sung J, Kim H and Chang T 1991 *J. Phys. Chem.* **95** 7121–4
[12] Veniaminov A V and Mahilny U V 2013 *Opt. Spectrosc.* **115** 906–30
[13] Paeng K, Swallen S F and Ediger M D 2011 *J. Am. Chem. Soc.* **133** 8444–7
[14] Teng C et al. 2012 *Macromolecules* **45** 6648–51
[15] Chang L P and Morawetz H 1987 *Macromolecules* **20** 428–31
[16] Liu C Y and Morawetz H 1988 *Macromolecules* **21** 515–8
[17] Brown H R and Wignall G D 1990 *Macromolecules* **23** 683–5
[18] McGrath K J and Roland C M 1992 *Macromolecules* **25** 1366–7
[19] McGrath K J, Roland C M and Weiss R G 1993 *Macromolecules* **26** 6127–8
[20] Tretinnikov O N and Zbankov R G 2004 *Macromolecules* **37** 3543–5
[21] Sillescu H 1999 *J. Non-Cryst. Solids* **243** 81–108
[22] Patkowski A et al. 2001 *Phys. Rev. E* **63** 061503

IOP Conf. Series: Journal of Physics: Conf. Series **1062** (2018) 012021 doi:10.1088/1742-6596/1062/1/012021

Optical properties of phosphate glass with CdSe quantum dots in terahertz frequency range

Demchenko P.S., Grigoriev R.O. Denisultanov A.Kh., Sedykh E.A., Onushchenko A.A.*, Khodzitsky M.K.

ITMO University, Saint-Petersburg, Russia

*NITIOM of S.I. Vavilov State Optical Institute, Saint-Petersburg, Russia

E-mail: `petr.s.demchenko@gmail.com`

Abstract. Optical properties and spectral characteristics of phosphate glasses with CdSe QDs were studied using time-domain terahertz spectroscopy in transmission mode in the frequency range of 0.2 - 1.0 THz. Concentration and size of QDs were varied by changing heat treatment conditions of initial glass samples containing components of the semiconductor phase. The change of refractive index and absorption coefficient at the high frequency edge of the investigated range at variation of secondary processing parameters was observed.

1. Introduction

Terahertz (THz) is the frequency range of the electromagnetic spectrum lying between the microwave band and the far infrared band. This frequency range isn't enough safe for human [1],[2] and it is often referred to as the last frontier for electromagnetic waves, since historically there has been comparatively little work observing the science and develop the applications of THz waves. Among all the applications of THz waves, the THz tunable metamaterial components, which may precisely control and manipulate THz waves, are essential for THz imaging systems and high-speed communication with THz rays [3] - [9]. Quantum dots (QDs) are nanoparticles with sizes in the range of nanometers, in which charge carriers are limited by the three coordinate axes, making the particles have discrete energy levels like an atom, that can be changed by varying the size of nanoparticles. Over recent years, semiconductor quantum dots (QDs) have attracted much attention of researchers due to the dependence of properties on size, and the creation of optoelectronic and photonic devices with tunable characteristics[10]. Currently, the QDs are being used used in the development of high-speed filters for signal processing, optical fiber amplifiers, low-threshold lasers, solar cells, etc[11]. The study of the properties of QDs in the terahertz frequency range is of special scientific interest[12]. In this experimental work, samples of barian-phosphate glasses grown in their volume QDs CdSe were studied for the development of new components and devices with tunable characteristics in THz frequency range.

2. Method of samples fabrication

The original glass was synthesized in silit furnace in a quartz crucible at a temperature range of 1380 - 1420 °C. Components of semiconductor compounds were introduced into the batch in excess of 100% in form of CdO and elemental phosphorus. Fusion was produced in heated graphite form. Annealing of the obtained casting at a temperature close to the glass transition temperature T_g was carried out. The original glass was transparent and had no absorption at wavelengths greater than 300 nm. The selection of the semiconductor phase was carried out as a result of secondary heat treatment at the isothermal conditions. In contrast to previous work [13], in which a QDs of CdSe were grown in a silicate matrix using the complicated two-stage heat treatments, in this paper we used a one-step heat treatment at a predetermined temperature in the range of 490 - 530 °C and a fixed time of isothermal heating for 3 hours, one sample at a temperature of 530 °C and a heating time of 7 hours and one sample at 550 °C and heating time 30 hours.

3. Experimental setup

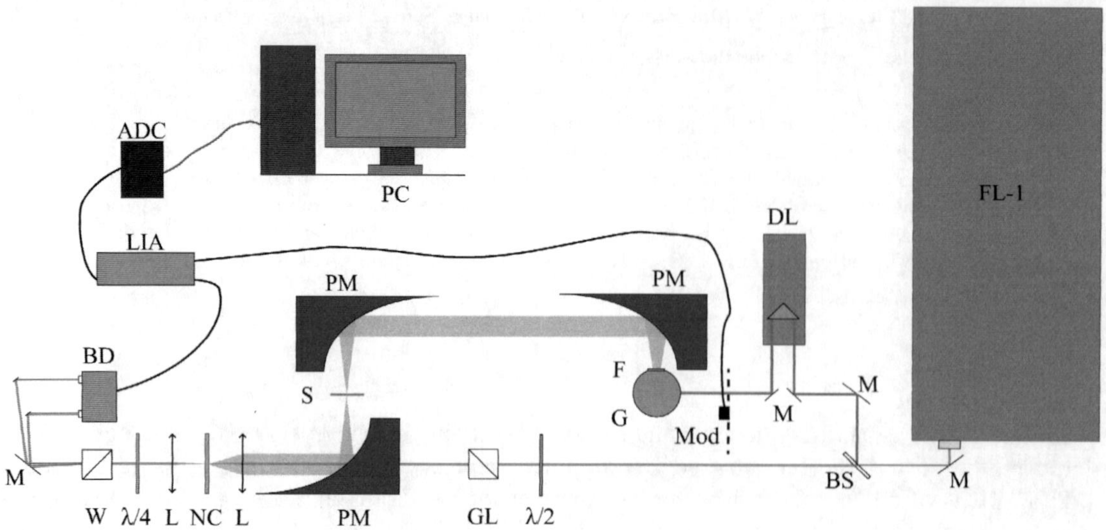

Figure 1. Scheme of pulsed broadband THz time-domain spectrometer (TDS). FL-1 — femtosecond infrared laser, M — mirror, BS — beam splitter, DL — delay line, G — generator (crystal InAs) THz radiation, PM — parabolic mirror, L — lens, F — Teflon filter, od — modulator, NC — nonlinear crystal CdTe, W — Wollaston prism, BD — balanced photodiodes, LIA — Lock-in-amplifier, ADC — analog-to-digital Converter, PC — personal computer

The optical properties and spectral characteristics of the samples were measured using time-domain terahertz spectrometer (Fig. 1). Infrared (IR) femtosecond laser beam is divided into a pump beam and a probe beam at 90% and 10% intensity, respectively. The pump beam passes through the delay line and falls onto the semiconductor indium arsenide (InAs) crystal surface. In this crystal, the Dember effect is observed: mobility of electrons in semiconductor is more than mobility of holes, thereby forming electric dipoles. The dipoles excite the radiation in terahertz frequency range. Next, the THz beam is focused into sample and passes through it and then incidents on the nonlinear crystal (NC) surface. The probe beam passes through: optomechanical modulator to remove acoustic noise and electromagnetic oscillations AC, $\lambda/2$ plane to rotate probe beam polarization on 45°, Glan prism to remove unwanted polarization. Then it falls into NC surface. The nonlinear crystal of cadmium tellurium is used to detect terahertz radiation due the Pockels effect. Without THz radiation the nonlinear crystal is isotropic for the IR probe beam. Otherwise, the crystal becomes anisotropic, the polarization of the probe beam becomes elliptical. Further the probe beam is collimated and it passes through $\lambda/4$. The changes of the degree of ellipticity of beam polarization are used in measuring process.

Further, the probe beam passes the Wollaston prism, after which it splits into two orthogonal components with vertical and horizontal polarization. These beams fall onto diodes of the balanced detector. The signal from the detector is analog signal of the difference of the beams intensities. THz spectrometer has the following parameters: the wavelength of femtosecond laser of 1046 nm, the IR pulse duration of 200 fs, the IR average power of $\geq$1 W; the THz pulse duration of 2.7 ps, the modulation frequency of 663 Hz, the frequency range from 0.2 to 0.8 THz, the dynamic range of $\geq$50 dB.

4. Results

Using the method of time-domain terahertz spectroscopy, the waveforms of terahertz pulses, which passed through the samples and air (Fig. 2) were obtained.

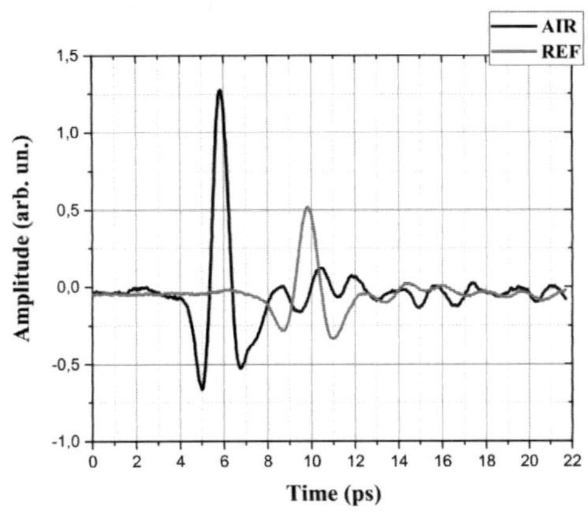

Figure 2. Waveform of terahertz pulses passed through the air (AIR - black line) and the reference sample (REF - red line).

The reference sample (REF) was without crystal phase CdSe QDs. The Fast Fourier transform (FFT) was used to obtain spectral characteristics and optical properties of the samples using magnitude ($A(f)$) and phase ($\phi(f)$). From these data we calculated the refractive index ($n_{re}(f)$), the permittivity ($\varepsilon_{re}(f)$) and the absorption coefficient ($\alpha(f)$) of the samples.

$$n_{re}(f) = 1 + \frac{c}{2\pi d}\frac{\phi_s(f) - \phi_r(f)}{f} \tag{1}$$

$$\alpha(f) = \frac{1}{d}\, ln\left(\frac{A_r(f)}{A_s(f)}\right)^2 \tag{2}$$

$$\varepsilon_{re}(f) = n_{re}^2 - \left(\frac{c}{4\pi}\frac{\alpha(f)}{f}\right)^2 \tag{3}$$

where d — is the sample thickness, c — the speed of light. The dispersions of refractive index, permittivity, absorption coefficient of the samples are shown in Fig. 3—5.

Also, if we make approximation that glass with QDs is effective medium, knowing the absorption coefficients of glasses with QDs and without them, it is possible to calculate concentration of QDs in the glass.

$$C_{CdSe}(f) = \frac{n_{re\ s}(f) - n_{re\ REF}(f)}{n_{re\ REF}(f) - n_{re\ CdSe}} \tag{4}$$

IOP Conf. Series: Journal of Physics: Conf. Series **1062** (2018) 012021 doi:10.1088/1742-6596/1062/1/012021

where $n_{re\ s}(f), n_{re\ REF}(f), n_{re\ CdSe} \approx 3.335$ — the refractive index of the sample, the reference sample and the bulk CdSe respectively. The dependence of CdSe concentration on the temperature of secondary processing for $f = 0.7345\ THz$ is shown in Fig. 6.

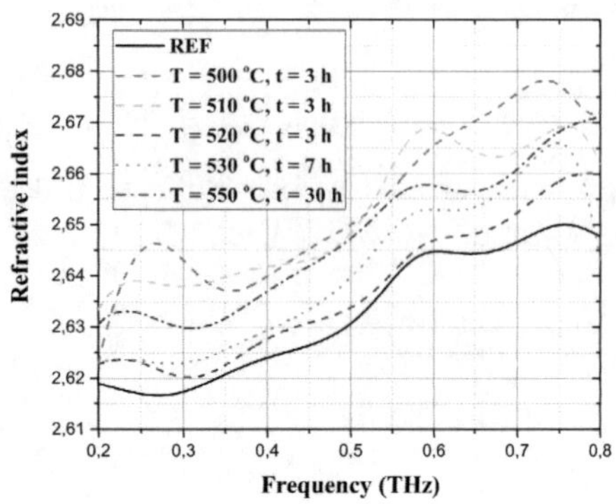

Figure 3. Dispersion of the refractive index of the investigated glasses

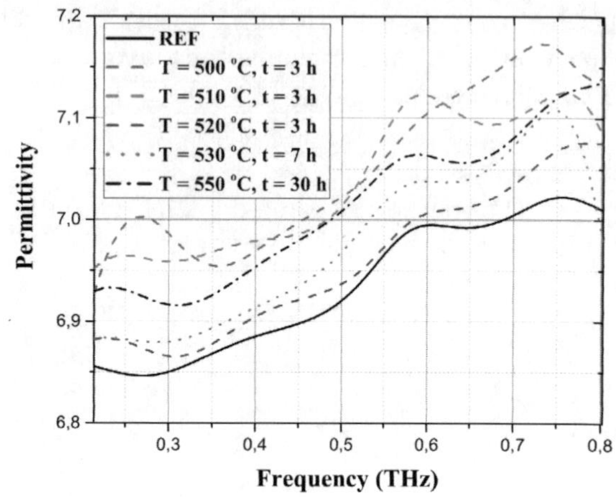

Figure 4. Permittivity dispersion of samples

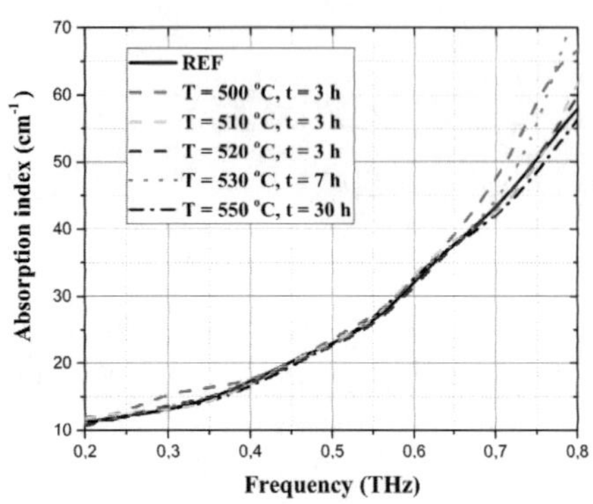

Figure 5. Dispersion of the absorption coefficient of the samples

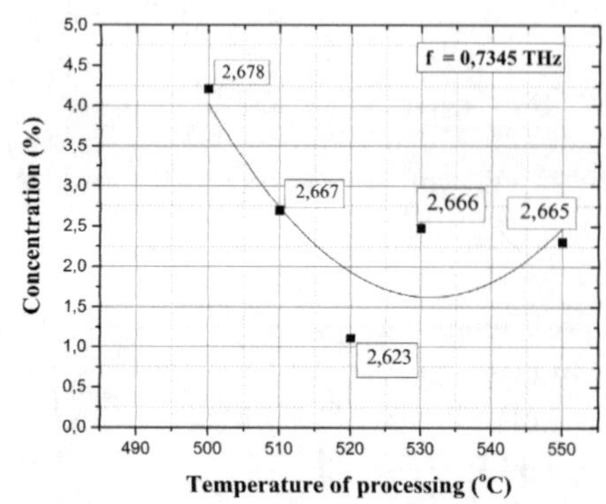

Figure 6. Dependence of concentration ratio on temperature of secondary processing of glass for $f = 0.7345\ THz$ with indicated refractive indices

5. Conclusion

We observed that CdSe QDs concentration varies from 4.21 to 1,11 % at the change of parameters of secondary treatment, that results in the increase of the refractive index of glass and the enhancement of the absorption coefficient. For the samples treated at the different temperatures but with the same exposure time, there is a linear dependence of the concentration on the temperature of secondary processing of glass. Based on this phenomenon it is possible to create terahertz component on the basis of glass with CdSe QDs, the characteristics of which can be controlled in the manufacturing process.

Acknowledgments
This work was financially supported by the Government of Russian Federation, Grant 074-U01.

References
[1] Borovkova M, Serebriakova M, Fedorov V, Sedykh E, Vaks V, Lichutin A, Salnikova A, Khodzitsky M 2017 *Biomedical Optics Express*, **8**(1), 273-280

[2] Duka M V, Dvoretskaya L N, Balbekin N S, Khodzitskii M K, Chivilikhin S A, Smolyanskaya O A 2014 *Quantum Electronics*, **44**(8), 707-712

[3] Denisultanov A Kh, Khodzitsky M K 2013 *Proceedings of SPIE - The International Society for Optical Engineering*, **8806**, 88062I

[4] Girich A, Khodzitsky M, Nedukh S, Tarapov S 2011 *NATO Science for Peace and Security Series B: Physics and Biophysics*, 159-164

[5] Tarapov S I, Khodzitsky M, Chernovtsev S V, Belozorov D, Merzlikin A M, Dorofeenko A V, Vinogradov A P, Inoue M, Granovsky A B 2010 *Physics of the Solid State*, **52**(7), 1427-1431

[6] Tarapov S I, Khodzitskiy M, Chernovtsev S V, Belosorov D, Merzlikin A M, Vinogradov A P, Granovsky A B 2009 *Solid State Phenomena*, **152-153**, 394-396

[7] Khodzitsky M K, Vozianova A V, Gill V V, Chernyadiev A V, Grebenchukov A N, Minin I V, Minin O V 2016 *Proceedings of SPIE - The International Society for Optical Engineering*, **9918**, 99182X

[8] Gurvitz E, Andronaki S, Vozianova A, Khodzitsky M 2015 *Optics Express*, **23**(20), 25738-25746

[9] Gill V V, Vozianova A V, Khodzitsky M K 2015 *Journal of Physics: Conference Series*, **643**(1), 012052

[10] Schmitt-Rink, Miller D and Chemla D 1987 *Phys. Rev. B* **35**, 8113-8125

[11] Alihosseini F et al. 2012 *6th Int. Symposium on Telecommunications (IST)* pp 48-51

[12] Hoffmann M et al. 2010 *Appl. Phys. Lett.* **97**, 231108

[13] Onushchenko P et al. 2015 *Proc. SPIE* **9502**

[14] Palik and Edward D 1998 Handbook of Optical Constants of Solids *Academic Press*

IOP Conf. Series: Journal of Physics: Conf. Series **1062** (2018) 012022 doi:10.1088/1742-6596/1062/1/012022

Chromene-3 and PM-567 bulk laser-active media tuning range

D S Tabakaev[1], E N Telminov[2,3], T A Solodova[2], E N Nikonova[2,3] and T N Kopylova[2]

[1] ITMO University, 49 Kronverkskiy prospect, St. Petersburg, 197101 Russia

[2] Siberian Physical-Technical Institute, 1 Novosobornaya square, Tomsk, 634050 Russia

[3] National Research Tomsk State University, 1 Lenina prospect, Tomsk, 634050 Russia

E-mail: dstabakaev@gmail.com

Abstract. In this paper we demonstrate first, to the best of our knowledge, experimental characteristics of tunable pulsed bulk laser on 3-(diethylamino)-7-imino-7H-[1]benzopyrano[3',2':3,4]pyrido[1,2-a]benzimidazol-6-carbonitril (Chromene-3) and 4,4-difluoro-1,3,5,7,8-pentamethyl-2,6-diethyl-4-bora-3a,4a-diaza-s-indacene (PM-567) dyes in polymethyl methacrylate (PMMA) – tuning range and optimal dye concentrations in polymer matrix, from lasing efficiency and tuning range point of view.

1. Introduction

Dye lasers as highly efficient sources of tunable coherent radiation are in high demand and this field shows rapid growth, basically, considering development and investigations of alternative to liquid dye solutions – bulk polymer lasing medium. Necessity of finding this alternative is obvious from cumbersomeness of liquid dye lasers, and, more important, from inherent to liquid solutions decrease of lasing efficiency at higher pumping power densities caused by appearance of reflection index temperature gradient [1]. Moreover, one can find bulk polymer lasing media much more cost-efficient, comparing to crystal-containing solid-state laser systems.

However, photostability of matrices and dye make bulk dye laser uncompetitive comparing with its liquid counterpart. Currently, this obstacle was overcome in virtue of high lasing efficiencies – up to 80% [2] - that makes possible achievement of demanded lasing power while pumping power density is reduced, that, without saying, means higher lasing source. It was a pivotal point in bulk dye laser development.

2. Experiment

Chromene-3 (CAS 52372-39-1) and PM-567 (CAS 131083-16-4) dyes and PMMA (Fig. 1) were chosen for lasing media creation, due to outstanding lasing properties of former [2,3], and lasing media with $2{,}5*10^{-3}$, 10^{-3}, $5*10^{-4}$, 10^{-4} Mole/l concentrations of dye were synthesized [4].

Transversal pumping geometry allows one to achieve higher lasing output energy, thus lasing media is pumped through the surface, created by grinding and polishing of cylinder's generator, which, without saying, is parallel to lasing direction. Hence, lasing as such appears only in thin parietal volume due to high dye absorption coefficients.

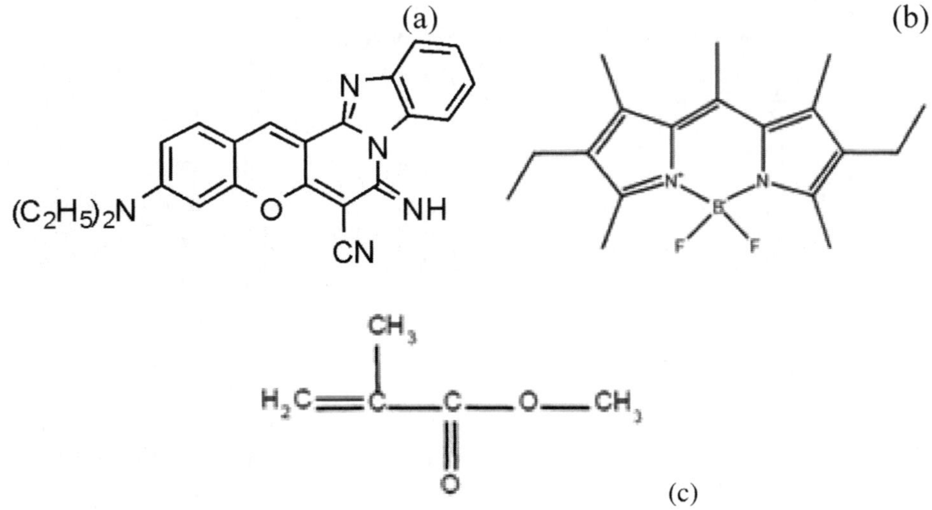

Figure 1. Molecular structures of: (a) – Chromene-3; (b) – PM-567; (c) – MMA.

Using these lasing media and selective transversal pumping cavity with grazing incidence diffraction grating (Fig. 2), lasing efficiency of lasing media with different dyes and concentrations were obtained.

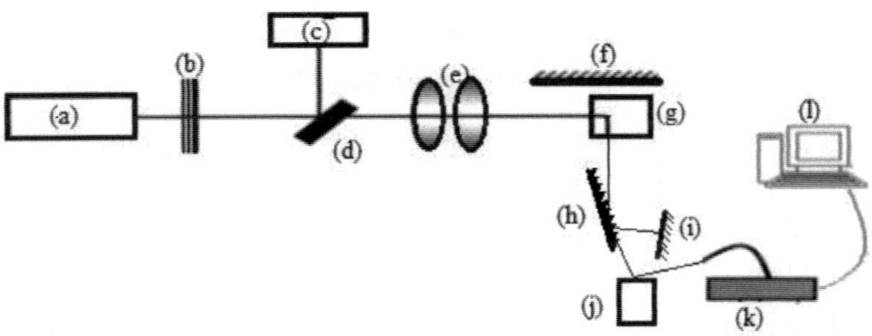

Figure 2. Scheme for tunable bulk dye laser prototype investigation: (a)- Nd:YAG^{3+}-laser; (b) – neutral density filters; (c) – Gentec EO ED-100A-UV energy meter; (d) – beam-splitter; (e) – cylindrical lence system; (f) – mirror; (g) –bulk lasing media; (h) – diffraciton grating; (i) – tuning mirror; (j) – Ophir NOVA II energy meter; (k) – Avantes spectrometer; (l) – PC.

3. Results and discussion

As a result of excitation of lasing media with 12 ns 532 nm Nd:YAG $^{3+}$ laser pulse, and rotating of the tuning mirror, lasing spectrum and tuning range for different dyes were obtained (Figure 3), except those one, which had PM-567 concentration lower than $5*10^{-4}$ Mole/l and Chromene-3 concentration lower than 10^{-4} Mole/l. It could be explained by appearance of lasing in longitudinal direction – coaxially to pumping (Figure 4), thus population inversion was consumed by these parasitic modes, related to this longitudinal “cavity”.

IOP Conf. Series: Journal of Physics: Conf. Series **1062** (2018) 012022 doi:10.1088/1742-6596/1062/1/012022

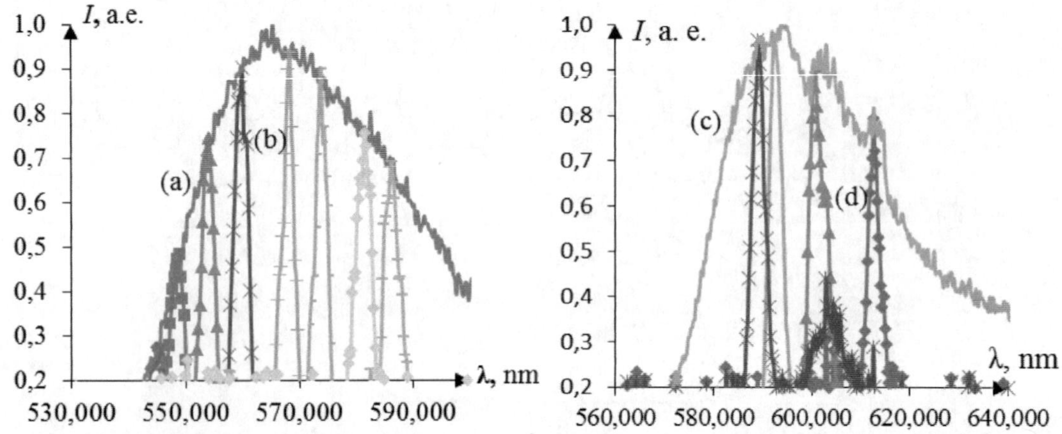

Figure 3. Normalized spectra: (a) – PM-567 10^{-3} Mole/l luminescence; (b) – PM-567 10^{-3} Mole/l lasing on different tuning angles; (c)- Chromene-3 $5*10^{-4}$ Mole/l luminescence; (d) –Chromene-3 $5*10^{-4}$ Mole/l lasing on different tuning angles.

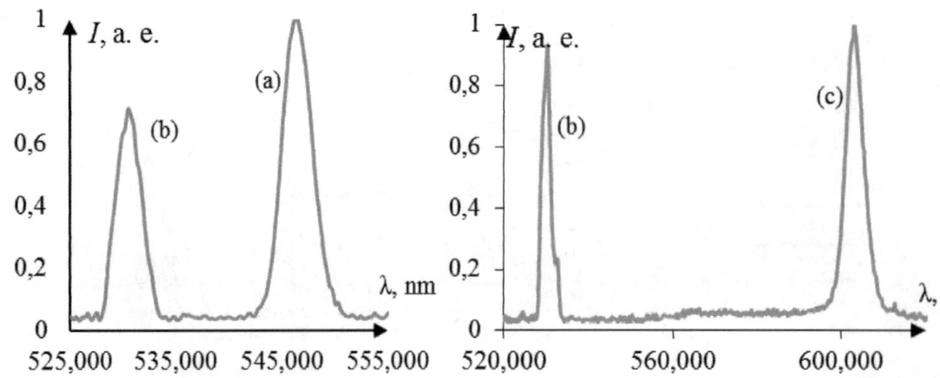

Figure 4. Normalised alsing spectrua : (a) – PM-567 10^{-4} Mole/l; (b) – Nd:YAG^{3+}; (c)- Chromene-3 10^{-4} Mole/l.

Analysing obtained dependencies, it is easy to find that tuning range of different dyes in such kind of prototype is close to 45 nm FWHM (full width at half maximum) and is possible for dye concentrations higher, than 10^{-4} Mole/l.

Under the same excitation conditions and using Ophir energy meter, lasing efficiency vs. pumping power density dependencies were obtained (Figure 5) and analysing it is clear that there is no decrease of lasing efficiency with rise of pumping power density, typical to liquid solutions of the same dyes. Moreover, to the best of our knowledge, these lasing efficiencies are the highest for bulk dye doped polymer matrices in selective cavities.

IOP Conf. Series: Journal of Physics: Conf. Series **1062** (2018) 012022 doi:10.1088/1742-6596/1062/1/012022

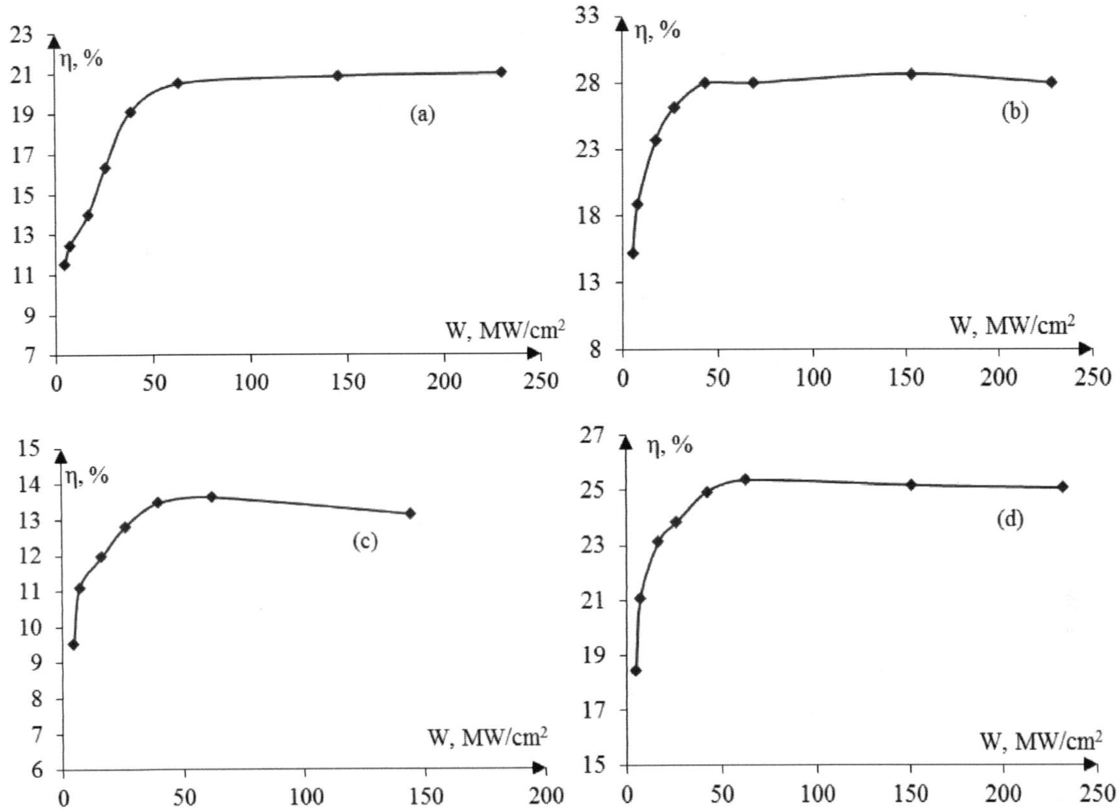

Figure 5. Lasing efficiency vs. pumping power density: (a) – PM-567 2,5*10^{-3} Mole/l; (b) – PM-567 10^{-3} Mole/l; (c) - Chromene-3 2,5*10^{-3} Mole/l; (d) - Chromene-3 10^{-3} Mole/l.

Hence, the highest achieved efficiency – up t o28% - is related to lasing media with concentration 10^{-3} Mole/l. This is explained by deeper propagation of pumping emission inside the lasing media, that entails smaller reabsorption and bigger lasing volume, comparing to lasing media with higher dye concentrations.

References

[1] Kopylov S M et al. 1991 Tunable dye lasers and their applications (Moscow: Radio I Svyaz)
[2] Tabakaev D S, Telminov E N, Solodova T A and Ponyavina E N 2013 *Russian Physics Journal* **56** pp 73–75.
[3] Kopylova T N, Tel'minov E N, Solodova T A, Ponyavina E N and Tabakaev D S 2015 *Russian Physics Journal* **58** pp 262–265
[4] Kopyova T N et al. 2012 *Izvestiya Vysshikh Uchebnykh Zavedenii. Fizika* **55** pp 32–37